FROM EARTH TO HERBALIST

An Earth-Conscious Guide to Medicinal Plants

GREGORY L. TILFORD

With a Foreword by Rosemary Gladstar
Illustrations by Nadja Cech Lindley

Mountain Press Publishing Company
Missoula, Montana
1998

Foreword © 1998 by Rosemary Gladstar
All photographs © 1998 by Gregory L. Tilford unless otherwise credited
Right photo on page 39 and inset photo on page 61 © 1998 by Paul Strauss
Illustrations © 1998 by Nadja Cech Lindley
Paintings on pages xiv and 216 © 1998 by Naomi A. Pelky
Herbicide-watch icon © 1998 by E. R. Jenne

Library of Congress Cataloging-in-Publication Data

Tilford, Gregory L.
    From earth to herbalist : an earth-conscious guide to medicinal
plants / Gregory L. Tilford ; with a foreword by Rosemary Gladstar ;
illustrations by Nadja Cech Lindley.
        p.   cm.
    Includes bibliographical references and index.
    ISBN 0-87842-372-9 (alk. paper)
    1. Medicinal plants.   I. Title.
RS164.T556   1998
615'.32—dc21   98-30002
                                                                CIP

PRINTED IN HONG KONG BY MANTEC PRODUCTION COMPANY

Mountain Press Publishing Company
P.O. Box 2399
Missoula, Montana 59806
(406) 728-1900

DISCLAIMER

❧

The sole intent of this book is to inform and entertain the reader through conveyance of recorded data and the personal experiences of the author and other herbalists. It reflects the theories, histories, and to no small degree, the author's personal opinions and speculations on the usefulness of wild plants.

This book is not intended to prescribe, recommend, or otherwise direct the reader in any level of medical self-care. Information it contains relating to the therapeutic uses of herbs is only intended to inform the reader on how other people have used botanical medicines. This book cannot in any way substitute for the care of a qualified health care practitioner or nutritional consultant, and the author strongly urges the reader to seek professional advice before using plant medicines or wild foods in any capacity. Herbal medicines may produce unanticipated or unwanted effects when combined with a preexisting health condition such as auto-immune deficiency or pregnancy.

When considering the use of wild plants as food or medicine, a strong measure of common sense is always indicated. Many of the plants in this book are very strong medicines and under certain circumstances can be toxic. Medicinal plants deserve a great deal of respect. One book or even a dozen books are insufficient to convey the knowledge one needs to safely and responsibly identify wild plants or use herbal medicines. Looking at photos and reading the text of one book is no substitute for proper training from a qualified instructor. With this in mind, the author strongly recommends hands-on field training with a qualified professional in the proper identification and use of wild plants before engaging in any level of harvest for the purposes of ingestion or therapeutic use.

The author, publisher, bookseller, and anyone else associated with the distribution of this book assume no liability for the actions of the reader. Use your common sense, and please consult a qualified instructor and/or health care practitioner *before* using plant foods or medicines.

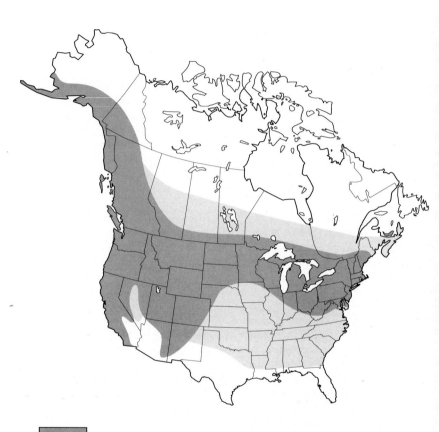

At least 75 percent of plants described in this book

At least 50 percent of plants described in this book

Less than 50 percent of plants described in this book

CONTENTS

~

FOREWORD

∽

People all over the world are alarmed at the destruction of the Amazonian rain forests. Are you aware that similar ecosystem ravagement is threatening the environment closer to home? Every day 2,400 acres of North American native habitat is destroyed and with it countless native trees and plants. According to some estimates more than 2,000 native plant species, including many medicinal plants, are in imminent danger of extinction in the United States. Habitat destruction through unsustainable logging and agricultural practices and sprawling urban development are the major causes of the dwindling plant populations, but the ever-increasing worldwide demand for herbal products also plays an unfortunate role.

When herbalism first resurfaced in the United States in the early 1970s as part of the "back to basics" movement, we herbalists entered the field wholeheartedly, harvesting a seemingly endless supply of wild American medicinal plants. As new herbalists, we zealously spread the word about using plant medicine by teaching classes, hosting gatherings, and making herbal medicines. We considered wildcrafting—the art of harvesting wild medicinals—essential to the practice of herbalism and often thought wildcrafted products superior to those made with their cultivated cousins. Small companies that later became industry giants sprouted during this fertile period, and their best-selling products were commonly based on wild-harvested plants.

Those humble early days of herbalism, rough and unconfined by rules, changed American medicine, as witnessed by the burgeoning interest in botanical medicine today. Herbal medicines are the fastest growing segment in pharmacies throughout the country, with more than a 50 percent growth rate in 1992. Consumers are expected to spend over $5 billion on herbal products by the year 2000.

While positive on one hand, the current herbal renaissance in American health care brings with it an ever-growing demand by consumers for herbal products. Where do all the plants used to make this vast amount of product come from? Until very recently, large-scale cultivation of medicinal herbs was rare. Almost all the plant materials used in botanical medicine came either from third world countries or from North American wild habitat. In parts of the world where traditional herbalism has remained intact for hundreds of years, finding the healing herbs in their native habitat now is often challenging.

China embarked in 1950 on an ambitious and successful program to integrate Traditional Chinese Medicine into the country's public health policy, and soon faced a severe shortage of wild populations of popular medicinal plants due to overharvesting. To compensate, China began a massive effort to cultivate medicinal plants and now has more than 1 million acres of medicinal plants under cultivation. But their wild resources remain in dire straits. India, the largest producer of medicinal plants in the world with over 2 million acres now under cultivation, also overharvested wild medicinal plants to the point of severe supply shortages. And British herbalist David Hoffmann recently informed me that in England, where much of our rich western tradition of herbalism stems, it is now illegal to pick herbs from the countryside due to overharvesting. In North America, goldenseal, American ginseng, and echinacea are among the commercially popular medicinal plants at risk of extinction in the wild due to habitat destruction and overharvesting.

In 1979 the first worldwide convention on plant conservation met in Thailand—its theme "Saving People by Saving Plants" and its all important mission to address the pressing need for global plant conservation. The much-quoted Chang Mai Declaration issued then heralded an overdue change in policy regarding plant awareness and conservation. Yet, twenty years later, the situation is worse, not better. Though governmental and organizational involvement is critical to the success of conservation efforts, the responsibility finally rests with the concerned individual. As Margaret Mead so eloquently stated, "Never doubt that a small group of concerned citizens can change the world. In fact, it's the only thing that ever has."

The ultimate question then becomes, "what can I do to ensure the continued survival—and abundance—of our precious plant resources?" Gregory Tilford helps answer that question in this book.

When I first met Gregory and Mary Tilford, they were homesteading in western Montana. Like many other people, they had migrated to the country seeking a simpler, more harmonious lifestyle. Living so close to nature, they developed an intimate relationship with the plants around them.

Gregory and Mary either organically grew or wildcrafted all the plants they used in their small, home-based herbal company, then processed, packaged, and sold the products they made. Unlike many herb companies, they maintained—and consequently never lost—a direct involvement with the plants each step of the way. They were already accomplished herbalists when I met them, and I was impressed not only with the depth of their knowledge but with their refined sensitivity to the plants they harvested and the environment in which the plants grew. Gregory's keen sense of observation and plying mind naturally lent themselves to his work as an herbalist.

Gregory and Mary were among the first to raise ecological questions pertaining to herbalists and the use of certain native species. Just as the herbal industry swelled to unprecedented size, Gregory published *The EcoHerbalist's Fieldbook*. The first book to voice the growing concern among the herbal community about our role in the diminishing populations of wild medicinal plants, it signaled an "in house" wake-up call.

Now Gregory offers *From Earth to Herbalist,* a deeply moving treatise to the plants, to guide us on the first steps toward sustainable use and preservation of healing herbs. The information contained within is practical, timely, nonpragmatic, and sometimes challenging as it invites us to reevaluate and question our role as herbalists. Without setting dogmatic rules or attempting to know all the answers, *From Earth to Herbalist* poses important questions about and offers sound suggestions for sustainable plant management. If we are going to use these wild resources, Gregory postulates, we must first understand plant and animal interdependence, our environmental impacts, sustainable alternatives and adjuncts, cultivation procedures, and sustainable wild harvesting methods. Gregory knowledgeably explores these topics for each of the native medicine plants covered in this book. He urges us to embrace the responsibility for plant preservation.

A breakthrough book in its own right, *From Earth to Herbalist* serves as a timely reminder for all of us who love and use plants. Gregory challenges us to reconsider our roles as herbalists, to go beyond health care consultant, medicine-maker, wildcrafter, and gardener/farmer to become earth-steward herbalists, protectors of the wild gardens. In doing so we gain sure footing to continue the work of plant conservation.

The issues of medicinal plant conservation among herbalists and herb users are many and complex, and the opinions among the lively herbal community itself differ as to what is or is not at risk. Nonetheless, the concern that supersedes all others, and that surfaces repeatedly in Gregory's writing, is for the welfare of the plants themselves. Frances Bacon, the English poet, once wrote, "one cannot pluck a flower without troubling a star." What then, if we destroy a species?

If we choose to use plants as medicine, we then become accountable for the health of the wild gardens. We begin a cocreative partnership with the plants, giving back what we receive—health, nourishment, beauty, and protection. We have reached a time in history when ignoring this relationship with the resources we use would be disastrous.

Herbalists have profoundly influenced the survival and resurgence of herbal medicine in this country, even dictating which plants are popular and commonly used by the herbal industry. It is time for us to have as profound an influence on the preservation of the plants we use, so that on a sunny afternoon many years from now, our grandchildren may take their grandchildren's children out to show them the plants their ancestors used, still growing in great abundance in the native wild gardens. It's a dream worth waking for.

Rosemary Gladstar
Founder of United Plant Savers

This fairy slipper *(Calypso bulbosa)* is one of the rewards of taking a closer look.

PREFACE

∾

Through innocent eyes, a young schoolboy in the late 1960s viewed the yellow flower not as a weed but as a unique creation, something beautiful and mysterious that turned its face to the sun—a brightly painted picnic bench for brilliant butterflies. Later the flower transformed into hundreds of tiny white fairies, borne away on the summer breeze.

Curious about the plant, the boy visited the public library and learned that it was edible and nutritious, and it may have been brought to North America on the *Mayflower* as a cure-all. Thrilled by his discovery, he shared his newfound wisdom with his third-grade friends. He boldly ate the plant's bitter leaves, enjoying the look of surprise on his friends' faces.

Since those early days of mine, I have come to know that fascinating flower as *Taraxacum* species, the common dandelion. Society has continually urged me to view dandelions from a different, more "mature" perspective. I am told time and again that dandelion is "a terrible weed" that should not be tolerated under any circumstances: it is evil, invasive, and serves no purpose but to make life miserable for gardeners and farmers. For years, I believed that was true, particularly during my teenage years when I worked a summer job as a gardener.

One day several years later, I noticed lush bunches of dandelion greens for sale in the produce section of a gourmet grocery market. Amazed that people would pay for the same weeds I had battled in my high school years, I bought some. I took them home, tossed them with lettuce and tomatoes, and ate them with a sharp lemon-herb dressing. Delicious! I gobbled a second serving, realizing that the curiosity of that little boy was still alive in my heart.

Today, I am an herbalist and a teacher, accepting with humility the natural offerings of well-being presented from the earth and sharing those gifts with others. I have learned that dandelions have powerful diuretic and liver-stimulating medicines and a vast array of minerals and vitamins—all of which can cleanse the blood and the body of waste and toxins. I have learned that dandelions contain glycosides, triterpenoids, choline, lecithin, inulin, and myriad other chemical compounds that scientists have shown possess medicinally active properties. I have learned how to make my own medicines from every part of the dandelion plant, using the leaves for a nourishing diuretic, digestive tonic, and eyewash; the roots for a potent liver stimulant and anti-arthritic remedy; and the flowers to help metabolize fat and provide relief from mild pain. I have even learned that the sticky, white latex that bleeds from the flower stem has been used for centuries as a remedy for warts.

Over the past twenty-five years, I have built on my experience with dandelions to employ the healing powers of hundreds of plants, each containing dozens or even hundreds of medicinal properties. As I continue to learn the countless lessons offered by the plants around my home, I find myself spending less time harvesting them and more time watching and learning from them. I am learning how plants provide food and shelter for insects and other animals that feed on their leaves and flowers; about how every organism in the forest reproduces and survives. I watch how the flowers of tall plants attract pollinators to their shorter neighbors. I am beginning to see how birds and animals serve as nature's farmers by eating and later passing the seeds of plants. I smell the pungent odor of such plants as yarrow and observe that their unique fragrance attracts certain types of insects while repelling others. I feel the sticky fruits of ripe shepherd's purse and realize that they capture small insects, which die and help replenish the soil at the base of the plant.

This book is about much more than how to use medicinal plants; it is about reconnecting with the living Earth. More than teaching how to use medicinal plants to serve human needs, it is about working with them to balance our health and Earth's health. It is about encouraging everyone to live responsibly on the planet, to tread lightly, and to realize that no one rides for free. It is about finding our way back to the place in our hearts where the inquisitive child waits, holding the key to undiscovered possibilities.

And perhaps it all starts with something as simple as tasting a dandelion.

ACKNOWLEDGMENTS

∾

First and foremost, I extend my loving arms to Mary, my soul mate and teacher. For the last fifteen years you have illuminated my path in life with your love and support. Words simply cannot express how much I love, thank, and respect you.

A warm thank you to herbalist Rosemary Gladstar, for writing the foreword for this book and for sharing so much of her wisdom, kindness, and warmth with me and others. Had it not been for your friendship and the hard work you have done as founder of United Plant Savers, I might not have found the special energies I needed to complete this book. It is a great honor to share the same path with you.

To Nadja Cech, thank you for your beautiful drawings. They convey the passion of a true artist and the love of a caring herbalist.

A hug to my special friend, Paul Strauss. Thank you for the great photos of a blue cohosh flower and ginseng berries. But more important, thank you for your landmark contributions to the future of wild plants. Your lifelong dedication to the 700 acres of Appalachian heaven you meticulously nurture is an inspiration to all of us who care about the fate of our planet. Keep up the good work!

To my mother, Naomi, who painted two beautiful pictures for this book: from the moment you brought me into this world, you have always painted a special picture of nature for me. Thank you for the color and depth you will always add to my life. I love you!

And, a bear hug thank you to all of the teachers, healers, hard workers, and visionaries who are dedicated to the healing efforts of United Plant Savers—you hold the guiding light into a healthy green future for plants and our planet.

Above the Medicine Tree —Naomi A. Pelky

EARTH MEDICINE

Herbalism is a holistic approach to health and healing that addresses not only the discomforts of disease but the concept of wellness. Today we use herbs in many different capacities, often as natural alternatives to over-the-counter and prescription medications. Wild cherry bark *(Prunus virginicum)*, for instance, is used in place of codeine to suppress a cough, and topical preparations of cayenne (*Capsicum* species) reduce the pain and inflammation of arthritic joints. Used in this context herbs can help relieve the discomforts of illness or injury, but like most medicines that are used in this capacity, they can only approach the symptoms of disease. The holistic healer sees disease as a symptom of an underlying "ill" that stems from deeper problems of the psyche and lifestyle. The focus of holistic healing is not to kill or cut out disease as it appears, but to prevent disease by reestablishing and maintaining a state of balance and rhythm between mind, body, spirit, and environment. It is an approach to healing based on believing that illness is the result of an imbalance or disruption spanning the entire physical and nonphysical being, not just the deficiencies and/or impairments of one or more organs or body systems.

Holistic philosophy is fundamentally broader than that of conventional Western medicine, which centers on intervention and suppression of disease symptoms. Western medicine is remarkable—almost miraculous—at bringing relief in times of crisis or when preventative measures can no longer be applied. And Western medicine has transformed herbs into phytopharmaceuticals—literally "plant drugs"—that save thousands of lives each day. About 40 percent of all prescription and over-the-counter pharmaceuticals come directly from plant-held compounds. Foxglove, as one example in thousands, is the unrefined basis for the important cardiac drug digitalis. Several species of nightshade are useful in the production of atropine, a pupilary dilator and cardiac sedative. However, while mainstream Western medicine works well at intervening with the symptoms of disease, a growing number of people are becoming dissatisfied with its approaches toward health maintenance. Many people are becoming weary of the politics, government regulations, and huge economic interests that are involved in the current administration of their health care.

The holistic practitioner uses plant medicines quite differently. Most plant drugs are derived from one or a few chemical compounds that have been isolated from the thousands that comprise a whole plant. The holistic healer uses a more complete representation of the plant, focusing not only on the most active chemical constituents, but also on the belief that there are less definable ways by which all elements of a plant's chemistry intermingle to make a complete medicine. Included in this belief are theories that the "life energies" of plants—the phenomena that separate living things from non-living things—somehow work in synergy with the life energies of the recipient to trigger a healthful response.

Instead of opting for a quick fix to the symptoms and discomforts of disease, effective holistic healing requires an investment of time to learn the fine details of a person's life and surroundings. From this, the holistic healer identifies the underlying causes of illness and devises long-term solutions conducive to the body's natural design. In other words, the holistic healer does not focus on the speculative prevention of illness, but on the maintenance of wellness. Virtually every type of holistic medicine stresses proper nutrition and exercise as the basis of good health. The body will naturally maintain and heal itself, but only if all elements of the collective whole are working together effectively and cooperatively. To accomplish this, the body must receive the nutritional, medicinal, and preventative components it needs to feed and support its interdependent functions—and the body needs a clean environment and healthy mind-set.

To maintain good health and well-being, the body sometimes requires a specialized source of stimulation or systemic support. That is where medicinal plants come in. They bridge the gap between what the body receives as part of complete nutrition and what it requires from time to time in supplemental support. When used in their proper holistic context, herbs help the body with what it is designed to do naturally: *stay healthy.*

By helping the body maintain and heal itself, herbs can serve as complementary adjuncts in a wide variety of therapies, including those that are not holistic. For example, in mainstream Western medicine, herbs can be useful for strengthening or moderating body functions that are influenced by drug therapies, or for potentiating the effects of the drugs themselves. In the patient who is using prescription anti-inflammatories in the symptomatic treatment of gout, herbs such as dandelion root (*Taraxacum* species), shepherd's purse *(Capsella bursa-pastoris)*, or yellow dock *(Rumex crispus)* may assist the liver and kidneys in eliminating excess uric acid (a primary cause of gout) from the body.

This is a book about the art, science, and philosophy of using medicinal plants in a way that complements nature. It provides a practical introduction to hundreds of North American plant species and their medicinal uses for plant enthusiasts of all levels who are interested in learning more about plant medicines. The fifty-two detailed plant profiles lead you through the basics of plant identification, medicinal actions, useful applications, ethical harvesting methods, and herbal medicine-making. But more importantly, you will learn new ways of sustaining these precious medicinal plants into the future. The profiles include suggested alternatives for each herb covered, instructions on how to replant them in their natural habitats, how to grow them in your garden, and how to minimize harvest-related damage to sensitive environments.

This book, though, is more than just another how-to manual about harvesting and using medicinal plants. It is about understanding the needs and natural roles of plants—extending the principles of holistic healing beyond the human body to include our species as one of many living components on that larger body we call earth. I do not condone commercial wildcrafting of any native plant that can be successfully cultivated. Nor do I support the use of any environmentally sensitive species, especially if resilient alternatives exist. I do encourage the earth-conscious (respecting the natural needs, sustainability, and design of each plant) harvest and use of most plants we commonly refer to as imported weeds, as well as the careful and limited use of native species that can withstand human impact.

From notes I have compiled throughout years of intensive field studies, I will introduce you to the daily lives of medicinal plants and the interdependence of plants and their habitats. From these pages, you will begin to learn the deepest level of herbal healing: using wild medicinal plants in balance with nature.

Whether you live in New York City, in the wilds of Alaska, or anywhere in between—and whether you ever choose to harvest a plant for self-care—my wish is to sow a seed of inquisitiveness and creativity in you that will transform the "weeds" in your life into healing allies.

The Herbalist's Connection with the Earth

When I reach into the soil to gather roots for formulating medicines, the pressures of society quickly melt away. And when I look beyond my needs, I begin to understand and address the needs of the natural world. I become interested in healing more than myself; I become interested in the health of the planet.

Such a perspective, which views humans not as the ruling entity on earth but as an interrelated part of a giant, living organism, has been called the Gaia—or living earth—philosophy. Many aboriginal cultures embrace this living earth philosophy. Healers in such societies are attuned spiritually, sensually, and intellectually to the healing powers of plants in ways that not only benefit humankind but also are harmonious with the earth's other creatures. The idea of modern-day medicine men and women may sound far-fetched, but recent findings and events are prompting even the most conservative scientists to take a second look at some of the ancient philosophies.

Old world diseases that scientists thought they conquered long ago are reappearing in areas where biodiversity has been diluted by the effects of

The forest beneath the trees—a Douglas fir seedling

pollution and overpopulation. Flesh-eating and drug-resistant strains of bacteria are evolving that defy our intellectual understanding. Many highly respected scientists now theorize that weather patterns are becoming more violent and deadly because of depletion of the earth's ozone layer and the effects of global warming. Some scientists even speculate that viruses may be the earth's antibiotic mechanism.

To change this rampant course of global imbalance, we must learn ways of healing that address our health as a part of a larger whole. We can begin the healing process by reconnecting with nature. As a source of healing that is recognized by every creature on earth, medicinal plants provide that connection. But we must look beyond the herb section of our favorite health food stores to consider the actual plants we are using. You can learn about the healing powers of plants whether you live near a wilderness area or a city park, or if the only plants you see grow from cracks in alleyways or sidewalks. All you have to do is realize and appreciate their existence.

That is only half the healing journey, though. We must also learn how to take care of what takes care of us.

The Casualties of a Green Revolution

In an ideal world, all wildcrafted herbs of commerce would be replaced with organically raised alternatives. But we do not live in an ideal world. Plant and animal species and natural resources are disappearing from our planet at an alarming and accelerating rate. By some estimates (United Plant Savers and World Wildlife Fund, 1997), as many as 20,000 plant species vanish from our planet each year. As I write this, 2,400 acres of North American wildlands are being lost to land development and agricultural activities each day, and a recent twenty-year study conducted by the Smithsonian Institution and fifteen other organizations indicates that at least one out of every eight plant species on Earth is now threatened with extinction. In North America alone, about 29 percent of 16,000 species are at risk. Changing this trend can begin only if and when we realize what is at stake and if we reevaluate what motivates our consumption of the earth's resources.

A large and rapidly expanding natural health care industry has recently developed, fueled by people eager to embrace alternative products, healing methods, and lifestyles. This could mean renewed hope for restoring and maintaining a healthy environment. It also brings about a new environmental crisis as we consume medicinal plants in greater volumes than they can sustain.

In the early 1970s, "natural products" meant alfalfa sprouts and purple potatoes sporting clods of rich, black soil from the backyard farms where they had been grown. Glass jars of freshly dried herbs were tucked onto shelves in a back corner of the health food store. Those who dared to remove a lid from one of those jars were rewarded with a unique, sometimes surprising aroma that stirred the imagination. The appeal of herbs in the 1970s didn't stem so much from scientific validation as from what they symbolized—a healthful, pure reconnection with the earth. Herbs were mysterious and promising, full of nature and reminiscent of simpler, healthier times.

Today, as I walk through huge trade shows that represent a snowballing, multibillion-dollar natural products industry, I see something quite different. As I try to avoid the pushy, muscle-bound models distributing chewing gum they say will boost my energy, increase my happiness, and push my libido into a state of heated rut, I see a vast cornucopia of free enterprise with little connection to Earth's medicine. Surrounded by echinacea shampoo, kava corn chips, and topical lotions that help melt away excess weight and dissolve cellulite, I envision the tons of wild-harvested herbs contained in the countless extracts, salves, and pills.

Many plant species are overharvested because they are misrepresented in the marketplace. Goldenseal *(Hydrastis canadensis),* for example, has been one of the most popular North American herbs of commerce for centuries, but it is also one of the most misused. Like many other herbs, goldenseal is sensationalized as a wondrous phytopharmaceutical silver bullet, a miracle medicine effective against virtually any microbe. The interest in goldenseal and other popular herbs comes from therapeutic success stories and promising scientific studies. Their mass consumer appeal, though, comes from marketing savvy and no small measure of consumer hysteria. Most plant medicines work within narrow parameters of therapeutic usefulness. Goldenseal is *not* a systemic antibiotic that will course through the body to fight infection like a shot of penicillin; instead, it works to inhibit only the microbes it directly contacts on the surface membranes of the digestive tract, eyes, or nasal passages.

It is bad enough to use something to death because we fail to take proactive measures to preserve it, but to wipe out a species because it is marketed as something it is not reaches beyond clever marketing to pure exploitation. While part of the problem stems from unethical practices of wildcrafting, or harvesting wild plants, many herbalists now exclude such wild plants as goldenseal, American ginseng *(Panax quinquefolius),* black cohosh *(Cimicifuga racemosa),* and osha *(Ligusticum canbyi)* from their personal medicine cabinets and wildcrafting efforts. This is a gesture, if not a preventative measure, against the plants' demise in the wild but only a Band-Aid treatment for a global disease.

Much of what drives this increased interest in herbs is born of solid scientific discovery and consumers' desires to take charge of their health and well-being. Certainly those are good things. But where do all the plants come from? Do the natural health care products really work the way we are led to believe? What makes herbs different from conventional medicines? What effect does our consumption of herbal products have on plant populations and habitat? For the answers, we must look beyond the hype and learn what herbs really are. Until we do, our herbal heritage will continue to disappear under the pressures of ignorance and greed.

Many medicinal plants are at risk of extinction because of factors totally unrelated to their use as medicines. Pipsissewa (*Chimaphila* species), for instance, is a "secret ingredient" in certain soft drinks. The habitat destruction associated with its harvest may be even more devastating than the loss of the plants themselves.

Besides consumer demands taking a serious toll on plants such as goldenseal and American ginseng, these and thousands of other plant species are also disappearing under the combined pressures of logging, grazing, mining, weed abatement programs, urban and suburban development, farming, pollution, recreational impact, and other human-centered activities.

Goldthread (*Coptis* species) is another inhabitant of old-growth forests that is quickly running out of places to live in the Pacific Northwest. Goldthread is not a particularly popular plant of commerce in North America—most of it comes from China, where it is widely cultivated—but the old-growth firs and cedars it lives beneath are vanishing at an alarming rate.

What Can You Do to Save Medicinal Plants?

On a community level, you can share what you learn from books like this one; you can educate people about the curative possibilities of plants growing around your home. Show your neighbor a medicinal plant along the roadside—such as mullein, dandelion, licorice, Saint John's wort, or milkweed—and raise the question of whether a weed is really just a weed. Tell people about the plight of goldenseal, noting that the greatest curative gifts of this plant may be undiscovered, and then show them how to use Oregon grape as a substitute.

On a personal level, you can accomplish even more. Join United Plant Savers (see "Resource Guide"), become involved with native plant clubs in your area, or participate in land restoration efforts. If you use herbs, learn as much as you can about the plants and the ecosystems in which they live. If the herbs you use are abundant in your area, are not at risk elsewhere, and you can confidently identify them, harvest them yourself while learning about *their* health care needs. Take time to evaluate and improve yourself as a consumer of our planet's resources, and learn how to give something back for everything you consume. You will be rewarded with a depth of holistic healing that cannot be attained at any herb store—I guarantee it.

Bioregional Herbalism: Getting to Know Your Own Backyard

Before harvesting wild plants, an herbalist must understand the ever-changing needs of an ecosystem. By focusing your wildcrafting efforts within the bioregion, or ecologically unique area, you call "home," you assume a greater sense of responsibility and become more involved with your environment's well-being. The greatest measure of ethical conduct comes from an intimate familiarity with the bioregion where you are harvesting. I call this approach bioregional herbalism, and it amounts to getting to know your backyard. You become connected, and you can't help but care.

Many plants are abundant in certain areas but rare in others. Learn to recognize and use plant species that are abundant in your home bioregion and resilient to the consequences of human actions. Each ecological niche is a small, unique part of the larger environmental picture. Learn the general status of the plants you are interested in from a global perspective. You may find that your role among certain plants around your home is strictly that of guardian.

You don't need to travel far to find what you need. Nature provides an incredible diversity of medicinal resources. This applies whether you live in the Adirondack Mountains or in Death Valley. For every medicinal plant that

There are at least three medicinal plants and several Douglas fir trees in this photo—do you see them?

grows in one bioregion, a functional counterpart exists in another. For example, to help heal a sunburn, the Adirondack herbalist might use self-heal *(Prunella vulgare)*, while in Death Valley the leaf juice of yucca *(Yucca schidigera)* will fit the bill. Even if you live in urban areas like New York City or Detroit, you will likely find the medicinal diversity you need growing from cracks in the sidewalks, in vacant lots, or in nearby woodlands (the problem, of course, will be finding plants that are free of toxic residues).

Learn the environmental issues of your area, and familiarize yourself with other people who have an interest in the wild lands where you will be working. Meet other herbalists, loggers, miners, developers, and people from environmentalist groups and government agencies. Ask about their plans for the land, and plan for the future by integrating your efforts in a positive way with theirs.

Contact your local county, state, or federal weed abatement officials and learn about weed control programs in your area before you harvest anything. Check your local "noxious weed list" and see if there is an opportunity to reduce the distribution of toxic herbicides by offering to hand-pull the plants for healing purposes. Many homeowners and farmers will welcome you on their land to gather such plants as dandelion, Saint John's wort, mullein, burdock, or countless others. By focusing your wildcrafting efforts on plants that people dislike, you will be giving a precious gift of healing back to the environment by reducing the use of toxic herbicides.

HARVESTING AND HANDLING HERBS IN THE FIELD

∾

Proper methods of harvest, transport, drying, storage, and medicine-making are essential in using herbs to their full potential, keeping waste at a minimum, and assuring natural sustainability of the plants. The information presented here applies to most of the herbs you are likely to harvest and use from wild areas of temperate North America and Europe. Bear in mind, however, that many plants require special handling, and every herb requires individualized attention to be used safely and effectively as a food or medicine. For specific information about individual plant species, please refer to the herb monographs that follow.

Guidelines for the Ethical Harvest of Wild Plants

If after familiarizing yourself with the character of the land, you choose to harvest your own herbs, assume responsibility for everything you take. Responsibility includes adopting a good set of wildcrafting ethics like these.

1. Never gather endangered or environmentally sensitive species. Ask yourself if your need justifies your harvest, before harvesting *any* species.

2. Make positive plant identification before you harvest. Gathering the wrong plants not only may be harmful to you but is wasteful and potentially damaging to an ecosystem as well.

3. Collect a small amount of plant material from several different stands to minimize any repercussions on the plant population. Never gather from a stand of plants to the extent that it makes a visible difference. Mathematical formulas for how many plants can be sustainably harvested do not apply in the natural realm; the only way to determine how much is okay is to study the ecology of your bioregion before you harvest. Always gather conservatively.

4. Never gather from a stand of plants until you have thoroughly investigated the health, welfare, sustainability, and interdependent elements of the entire area. Is another healthier stand nearby? Is the stand large enough to sustain the stress of your gathering? Do you know the

ecosystem well enough to harvest plants in a manner that is conducive to its good health? Are there good trails, or will your incursion into the area cause damage? What may seem a lush, healthy stand of plants could be a biocommunity in trouble. Walk, look, listen, feel, and—most important—learn as much as you can about the ways of the land, before you harvest.

5. Never gather more than you can conservatively use. Many herbs lose their potency soon after harvest. Unless you plan to make alcohol tinctures, expect a shelf life of six months to a year for properly dried and stored herbs.

6. Always obtain permission before entering private lands, even if you think you are doing the landowner a favor by pulling a few weeds. Check for permit requirements with your local land management agencies before gathering herbs from public lands.

7. Maintain a journal of harvesting site records and observations, and keep a detailed log for each wildcrafting venture (more on this later). From your records, you can monitor your long-term effects while enriching your natural awareness.

8. Learn the typical elevation, habitat, soil composition, exposure, and climatic requirements for each plant you wish to harvest, then harvest from areas matching those criteria. Small, isolated, stands of plants that have adapted to an unusual set of environmental circumstances are particularly vulnerable to human interference; use them for learning, not for making medicine.

9. Watch for opportunities to salvage or transplant plants from areas scheduled for large-scale development activities, such as trail or road construction projects, logging, or subdivision. Check if your area has native plant restoration projects under way in which you might help native species survive by "weeding out" invasive, non-native species.

10. Locate pristine wild areas that show no signs of human encroachment and contain healthy stands of the herbs you might need. Do not harvest from these areas. Instead, use them as control models to comparatively monitor the effects of your harvest and stewardship efforts at sites where you *will* be harvesting. The goal: to match the health of your harvest sites with the natural health of your control models.

11. Never gather from stands that are wilted or otherwise unhealthy. Sick plants may have altered chemistries, or they may be covered with poison.

12. Always use the correct tools and wear the right clothing to get the job done. Digging long taproots from a steep, rocky hillside with a soup

Perhaps the greatest level of human healing offered through collecting your own herbs is in learning where to step next. This delicate colony of false lily of the valley *(Maianthenum dilatatum)* would likely suffer if walked upon.

spoon while wearing penny loafers will result in physical trauma not only to the herbalist but also to the hillside.

13. When you enter a wild area, be as quiet as possible and plan every footstep. Stop frequently to look, listen, and feel the earth. Don't forget to look behind you to monitor any damage, such as soil compression, displaced rocks, or crushed seedlings, you may be causing while en route to "that special patch." Here's good rule for monitoring soil compression: if soft-soled shoes or bare feet create indentations that are more than ¼ inch deep and that don't rebound quickly, you are compacting the soil. The best place to gather is from stands with easy access over firm soil.

14. When gathering from a stand of plants, work at a slow, deliberate pace. Harvest carefully. Don't attack the plants—work like a loving caretaker.

15. Remain obscure during your harvest and field studies. If you attract attention to a stand of plants, it may result in added stress from curious

onlookers. A good rule is always to focus your efforts in areas where you will remain out of view from other people.

16. Good wildcrafting ethics extend beyond your work in the field. Always use proper processing and storage methods for the herbs you gather. Spoilage means waste. Take care of the plants, all the way from earth to herbalist.

17. Never reseed a stand of plants beyond the natural density that existed before your harvest. A human-engineered overpopulation of plants can be as devastating to an ecosystem as careless overharvesting. Observe over several years the bioregion from which you are harvesting, noting how each species reseeds itself. Then, try to replicate what nature would do in your absence. Many plants compete with one another. Propagating an overabundance of a species may satisfy your conscience but be a death sentence for dozens of plants and interdependent organisms.

18. Accept your weaknesses, identify your chronic mistakes, strive to deepen your awareness, then continue this list with your own set of wildcrafting guidelines.

Ethical guidelines are essential in maintaining a basis of moral and practical conduct in the field—they help us learn how to begin the process of learning kinder and more respectful ways of coexisting with an ecosystem. However, in addition to following a good set of wildcrafting ethics, we must make conscious and continuous efforts to understand how an ecosystem naturally operates.

To harvest plants in a sustainable manner that is truly respectful of natural needs, the wildcrafter must first realize that ethics exist only in the minds of humans. To do something ethical means to conform to a preconceived set of principles that the human species accepts as the correct thing to do. In other words, we are adhering to a self-imposed rule. The problem is, nature does not recognize humanity's rules. This means we must learn to look beyond ethical guidelines (or "best management practices") to see exactly what makes an ecosystem tick, both with and without human influences. Unless we do, good ethics can only serve to justify our exploits.

To illustrate my point, let's explore one "ethical guideline" that many wildcrafters (people who gather things from the wild) use to determine and justify the number of plants they take from the land: Take three and leave seven of every ten plants. This simple, easy-to-follow guideline sounds reasonable, but if we consider a few of the factors that may contribute to the effects of our harvest, it becomes clear just how devastating this approach can be. What about the next wildcrafter who comes along to harvest two of the seven plants that remain from our harvest, or the development project that is slated for the area next year? And what about natural events (such as grass-

hopper infestations, drought, changing forage patterns of resident animals, and forest fires) that may contribute to the overall effect of our harvest? Many of these things are difficult or impossible to foresee. In the end, the extent of the damage we cause in wild areas is not gauged by the rules we obey or disobey, but by the combined extent of our awareness, concern, and ethical behavior.

Keeping a Field Journal

Maintaining a field journal is a critical component in the ethical harvest of wild plants. By recording your observations and activities, you can keep the long-term effects of your harvesting incursions in check, at the same time learning how an ecosystem works from season to season. Review your notes from previous visits to a harvest area so you can track natural trends. Most important, you will be learning the rhythms, flows, and balances of the ecosystem.

The record-keeping essentials presented here are designed to help you develop a personalized monitoring system and field diary. I have listed the bare necessities of what I regard as good note keeping. Tailor your own system to what works for you and your bioregion. It is impossible to be too detailed in your note keeping.

Record-Keeping Essentials

Harvest Site

- Site location
- Dates of first visit and each subsequent visit
- Botanical name of each plant you intend to harvest from the site
- General characteristics of the site. Examples: "open woodland with scattered trees and small shrubs"; "riparian thicket with predominant brambles"; "manmade flood dike with lots of introduced weeds"
- Proximity to roadways, power lines, or other easements
- Investigative notes about possible previous herbicide use, future human encroachment, and so on
- Weather conditions during each visit
- Detailed inventory of every living organism you see during your initial visit, and updates during subsequent visits. Example: "Visit #3 on 8/8/97: the wild turkey that I saw during my last two visits was not here. A plump golden eagle nests on a rocky bluff nearby."
- Any signs of animal or insect activity, such as tracks, abandoned nests, droppings, or nibbled leaf stalks
- Inventory of plants you can identify, and a count of the ones you can't. Include notes about their overall health, and make notes of any changes you see over time.

Harvest Activities

- Species, plant parts collected (roots, leaves, flowers, and so on), and quantities of each herb harvested
- Specific location within the site area where the herbs were collected. Example: "Gathered from the southern edge of the large patch on the west side of the creek."
- Methods used to harvest the herbs: how you removed the plant parts you collected (dug, pulled, cut, plucked); how you cared for them afterward ("made into fresh plant tincture"; "cut and dried"); and notes about your successes and failures ("roots began to mold before thoroughly dry"; "strong, dark tincture after two-weeks' maceration")
- Notes and observations about visible effects your visit to the site may have caused and about any special observations you had during your visit. Familiarity with minute details of the ecosystem leads to less human damage and interference.

People aren't the only ones who find wild plants useful. What do you suppose has been eating this arrowleaf balsamroot?

Basic Methods of Harvest and Proper Handling

Harvesting Flowers and Foliage

Harvest most leaves, flowers, and other aboveground plant parts during the early stages of a plant's reproductive cycle, when they will contain the highest diversity and concentration of active chemical compounds. Gather leaves shortly before the plant blooms, and gather flowers just as their buds are beginning to open. When you will be gathering the upper leaves and flowers together, as with arnica, harvest as soon as the flowers begin to open.

Never harvest leaves and flowers when they are wet, and always allow them plenty of air circulation from the time they are cut to the time they are processed into medicines. Moisture is the enemy; if the herb does not remain dry, it very likely will develop mold and go to waste. Give extra care to plants (such as angelica, bugleweed, or chickweed) harvested from moist environments, as they usually have a higher water content than dryland plants (such as bee balm, arrowleaf balsamroot, or yarrow). Carry the herbs you harvest in mesh sacks or ventilated paper bags. Unless compost is your goal, never put freshly harvested herbs into anything plastic.

When harvesting leaves or flowers, pluck a selected few from each plant, taking care to avoid a visible effect on the stand where you are gathering. If you can see the results of your harvest, you have taken too much. Leaves are the solar-collectors of a plant, responsible for producing chlorophyll and for converting light energy into plant energy. The flowers are the plant's reproductive organs. If the plants you are harvesting reproduce annually, leave the reproductive capacity of the stand intact for self-seeding. Learn the reproductive requirements of the specific plants you are harvesting *before* setting off on a wildcrafting expedition. For instance, some perennials reproduce exclusively by seed, while others reproduce from roots, crown division, or by a combination of reproductive mechanisms. By knowing the plants' habits before entering the field, you will be prepared to meet their needs.

Harvesting
Arnica cordifolia

If your harvest involves cutting stems, use sharp pruning shears and always cut the stem at an angle, which helps the cut heal quickly and prevents harmful fungi, bacteria, and other microbes from entering the plant's internal tissues.

After you have harvested the quantities you need, process them as soon as possible.

Harvesting Fruits, Bark, or Small Limbs

When harvesting fruits (such as rose hips, hawthorn berries, or elderberries), the foremost considerations should be how important they are to resident wildlife, and how the mother plant reproduces. Be aware of insects and animals that need the fruits more than you do. If you see any evidence of forage or insect infestation, move to another stand of plants. Always gather from portions of the plants that are within easy reach. Don't climb trees or use special implements to gather fruits; leave the hard-to-get fruits alone, and be respectful of nesting birds or other wildlife that may use the plants for cover. Pluck selectively and conservatively, and move frequently from site to site.

Most fruits are collected when they are fully ripe. Berries need to be handled the same way as leaves and flowers. Help air circulate around them by stirring them frequently until you get them home, then spread them out on butcher paper or a nonmetallic screen until you can finish processing them.

Bark, such as willow, poplar, or red osier dogwood *(Cornus stolonifera)*, is usually best if harvested in early spring, when the leaves of the plant are just beginning to bud and the plant's inner bark, or cambium, is rich with various compounds. You need the green inner cambium, not the dead outer bark of the plant.

There are two methods for harvesting medicinal barks. The most common method is to use a sharp knife to strip some of the bark from one side of a few small limbs of the plant, shrub, or tree. With this method, you must be careful to assure that the limbs do not die as a result of your efforts. Never cut completely around the circumference of a limb or trunk. The cambium is the plant's circulatory system, and if vital nutrients cannot be carried to the limbs above the cut the limb or entire tree will likely die. The injury resulting from your harvest also might allow harmful bacteria or fungi to enter the inner tissues of the tree, leading to infection that can stunt growth, kill the limb, or even kill the whole tree or shrub. To help prevent this, you may apply a commercially prepared pruning compound to the wound, or better yet, seal the wound with pitch from a neighboring tree. If you intend to strip bark, do so *very* conservatively, taking a narrow strip from each of a few small branches. Monitor the effect over time, watching how quickly and completely the injuries heal. You might find that this kind of harvest can be sustained only once every few years.

The second, and better, method of gathering bark is more considerate of the host plant. A select few of the newest limb ends are snipped off with pruning clippers and are later chopped into small pieces for further processing. With this method, the key to getting plant material of optimum medicinal potency is to harvest as early in the growth season as possible—just as the new limbs begin aggressively to shoot away from the plant. Arguably, these limb shoots might not be as medicinally potent as the cambium of the larger limbs and trunk of the plant. In my experience, this method fills the medicine bill just fine, without scarring the plant for years to follow. Remember, though, that young, spring shoots of many tree and shrub species are an important source of food for wildlife. The young shoots also produce the catkins, flowers, or other reproductive parts of the plant, so be careful to leave plenty for the plant's reproductive activities.

Harvesting Roots, Rhizomes, and Stolons

The harvest of roots, rhizomes, or stolons often involves the death of plants. Therefore, spend some extra time learning the natural roles of the plants you want to harvest and the effects that will likely result from their absence. If you gather such plants as lomatium or arrowleaf balsamroot from loose soil on a steep hillside, you must consider the plants' role in preventing erosion. If you gather taproots from an ecological niche where the plants you are taking represent the primary source of forage and small animal cover, you must evaluate the effects of harvest on dependent wildlife. You need to know as much as possible about the ecology of the area where you are harvesting from before you begin to harvest, especially if your harvest results in permanent change.

Here are a few guidelines for harvesting any kind of root. First, spend a few bucks and get the right equipment to do the job. Don't go into a stand of lomatium (*Lomatium dissectum*) with just a hand trowel—you will need at least a hand pick and a good, strong shovel. Wear a pair of lug-soled boots, and don't forget gloves. The proper equipment will result in less personal wear and tear as well as less wear and tear on the land.

For general taproot digging, use a narrow-bladed ditch spade, one with a rounded tip and preferably a fiberglass or metal handle that won't break from root-digging abuse. This type of shovel penetrates hardpan soils and easily reaches the bottom ends of long taproots in a manner that is less obtrusive than a typical broad-bladed, all-purpose shovel. You will also need a good hand trowel and some form of hand pick—perhaps a pointed mason's hammer—so you can selectively clear away soil and rocks from the roots you want without damaging them or the roots of surrounding vegetation.

You will want to dig most taproots in fall, after the aboveground parts have died back and the plant is either finished or in the process of distributing its seeds. That is when the most concentrated diversity of polysaccharides, volatile oils, resins, and other medicinally active components are in the roots. With many plants, this is the *only* time when certain desirable constituents are in the roots. Echinacea root, for example, contains a full array of active constituents only after the plant goes dormant for the winter. Some exceptions to late fall harvest include second-year biennials, such as burdock (*Arctium* species) and mullein, which are at the end of their life cycle by the end of summer. The roots of these plants should be harvested no later than mid-August, or better yet, gather only the fall roots of first-year plants.

If you are collecting rhizomes or stolons, you will find a hand-held, clawlike garden cultivator or a manure fork handy for pulling them up. Rhizomes are stems that extend through the soil horizontally, such as nettle roots (*Urtica* species). Stolons are stems that crawl horizontally and produce roots from above ground, such as strawberry runners (*Fragaria* species). Some plants, such as Oregon grape (*Berberis* species) and Uva-ursi (*Arctostaphylos uva-ursi*), may produce both rhizomes and stolons. All tend to be strong and woody. They are best harvested by pulling the aboveground plant until you meet resistance. Then cut the rhizome or stolon with pruning shears.

Aside from these special techniques, most of what applies to the harvest of leaves and flowers is the same for roots. If you are gathering roots in wet or consistently moist environments, take extra care to ensure that they will not mold after harvest. Dryland roots, such as arrowleaf balsamroot (*Balsamorrhiza sagittata*) or lomatium (*Lomatium dissectum*), are better about staying fresh and medicinally viable. Most dryland roots can be stored in paper bags (with the tops left wide open) for several days before processing.

Here's the ethical bottom line to harvesting most roots: Once you dig it out of the earth, it is gone forever. No matter how successful your reseeding efforts, you have altered an intricately balanced piece of the planet. Dig roots slowly and considerately. Instead of ripping them from the earth, remove them with the careful deliberation of a holistically minded surgeon. Give thanks for what you have taken, and monitor the long-term effects of your harvest.

Gathering Seeds

Gather seeds from most plants during late summer or early fall, after they have ripened and dried but are still on the plant.

Seeds can be collected with a minimum of environmental damage, provided they are collected conservatively. Many wild plants have poor germination rates or long periods of seed dormancy. The seeds of some species must remain under or on top of the soil for several months, or even years, before germination. Some will not germinate unless they first pass through the digestive tract of an animal, and others require cold temperatures, hot temperatures (even fire), or direct exposure to sunlight to break their dormancy. While they wait, they may be vulnerable to hungry wildlife, the composting effects of fungi and bacteria, environmental changes, and other variables that may prohibit their success. This explains why you might see an apparent overabundance of seeds on a single plant. Such appearances alone are not accurate indicators of a plant's reproductive prowess. Even dense populations of the mother plant are not always a strong indicator of successful reproduction. Some plants, such as mullein (*Verbascum thapsis*) or fireweed (*Epilobium angustifolia)*, will grow in profusion for a few years, and then having served their specialized natural roles, they all but vanish.

If you plan to gather seeds, learn something about the germination characteristics and growth cycles of the species from which you plan to harvest. Study the plants to find out if you will be competing with dependent organisms, and gather the seeds you need conservatively, from several different areas in your bioregion.

Store seeds in plastic bags in a cool, dry, dark place. Following these guidelines, most seeds will remain viable for several years, but some are vulnerable to oxidation and will literally turn to dust within a year of harvest. To estimate the shelf life of a seed, examine the structure of its outer surface. Seeds that are soft, papery, or that crumble easily when rubbed between your fingers—such as those of angelica, shepherd's purse, or cow parsnip—are likely to be short-lived. Hard, stonelike seeds—such as those of cherries, elderberry (*Sambucus* species), and Oregon grape—are generally less pervious to their environment and will remain viable much longer.

THE BASICS OF MAKING
HERBAL PREPARATIONS

∾

The way you use an herb depends on a holistic assessment of the circumstances at hand and the overall nature of the person who will be using it. Once you have determined the origins of the symptoms, you can select the medicinal actions to best confront the problem. This will lead the herbalist to several choices from which he or she can choose one or more herbs that have a specific affinity to the body systems involved. Choosing the correct herb is critical to the effective treatment of disease and also to conserving botanical resources. The ability to make such determinations comes only from experience. If you are unfamiliar with the symptoms, causes, or nature of a medical problem, or if you are uncertain about which herbs may be appropriate or how they should be used, consult a qualified health care practitioner. To use an herb effectively and responsibly means that it should be used only in its therapeutic capacity and in a form or preparation suited to its medicinal attributes.

The characteristics of a particular herb's physical and chemical structures determine how it should be prepared and administered. Many herbs, for instance, are highly water soluble: if you pour hot water over them, their medicinal constituents quickly and completely "infuse" into the water, making their chemical attributes freely available for the body to absorb. Other herbs are poorly water soluble: if you pour hot water over them to make an infusion, little or nothing is drawn from the plant's tissues.

This chapter will introduce you to the basics of making herbal preparations. Specific information about which of those preparation methods is appropriate for the herbs described in this book can be found in the individual plant monographs.

Compared with other books about herbal medicines, the guidelines I am offering are fairly loose. Exact measurements and strict procedures are necessary measures of quality control in the commercial production of herbal products, but the self-reliant herbalist usually does not have to be overly precise. As long as you use enough herb and a strong enough solvent in the correct type of preparation, you will be rewarded with a good, strong, herbal medicine. For the person who is more comfortable with carefully spelled-out directions, I have included exact tincture-making formulas for each herb.

Dried Bulk Herbs

Dried bulk herbs are handy because they allow for diversity in preparation and application. The drawbacks are their limited shelf life and that some plants lose much of their medicinal potency and diversity when they are dried. Roots generally will keep two to three times longer than dried leaf material, but don't stock up on more than you can use in a year. Dried leaves and flowers break down even faster—plan on harvesting a six- to twelve-month supply.

The first step toward assuring that your dried herbs remain fresh and viable is to handle them properly in the field. Herbs that develop mold before you can get them home probably will spoil before they dry.

The second step is to properly care for the herb throughout the drying process. Spread your harvest loosely on butcher paper, an old bedsheet, or a plastic screen—anything but metal. Metal surfaces may have oxide residues that can speed the chemical breakdown of herbs or even change the herbal chemistries. Leaves and flowers should be left whole (as much as possible) throughout the drying process and until they are used. Small roots can be dried whole, but large and juicy ones should be cut into lengthwise quarters to allow them to dry more quickly and completely. Avoid chopping or grinding your herbs until you plan to use them. Powdered, chopped, or ground herbs oxidize and lose their potency much faster than those left in large pieces.

It's okay to sun dry herbs for a few hours, especially if they are damp. But don't leave them in the sun beyond that because ultraviolet rays speed the oxidation process. Ideally, herbs should be air dried out of direct sunlight, at temperatures between around 80 and 95 degrees Fahrenheit, at humidity levels between 20 and 50 percent. However, if these conditions cannot be met, it's better to dry your herbs more slowly (at a lower temperature or higher humidity) than to risk damaging them with an artificial heat source. Electric food dryers, unless they are expensive top-of-the-line models designed for even heat distribution and controllable temperature, are usually too hot. Space heaters are dangerous and can be too hot as well. Too much heat will destroy medicinal compounds, and too little will encourage mold. Any dark, well-ventilated area in the house or garage will do. Stir the herbs once or twice a day in warm conditions and three or four times a day if the area is kept at 70 degrees or less. This will speed the drying process and prevent moisture accumulations.

Your herbs are completely dry when they feel and *sound* completely dry. Properly dried leaves crackle and crumble when you squeeze them, but their color and aromatic qualities will be intact. Dried roots should feel completely dry and either snap when you bend them or have a resilient but woody consistency. Be absolutely certain that your herbs are completely dry before you store them. Otherwise, they will compost.

Store your herbs in plastic bags or airtight containers out of sunlight. Check them from time to time. Do they still smell fresh? Do they look the same? How do they taste? Any changes indicate that the herbs are getting old and are losing their potency. When this begins, do something with the herb right away. Make a tincture from it or give it to someone who needs it. And remember to harvest less next time.

Fresh Herbs

Fresh herbs—clipped, dug, or plucked directly from the earth—are rich in nutrients and medicinal constituents. When used in their whole, "raw" form, fresh herbs are beyond compare for medicinal potency. Some herbs, however, are too strong when used fresh, and others require at least partial drying before they will impart their active constituents into a useful medium. Cascara sagrada bark, for example, is an herbal laxative that is far too strong if used fresh; it should be dried for at least a full year to mellow it before use. Some herbs make better preparations when they are dried first. Uva-ursi is a good example. The stiff, leathery leaves are almost impervious to water, and even a strong alcohol solvent has a hard time penetrating them to pull out the active ingredients. For this herb, drying the leaves makes their cell structures more vulnerable to hot water, alcohol, and other solvents.

Some plants must be used or processed when they are field fresh, because they begin losing their medicinal potency as soon as they begin to die. Cleavers and chickweed, for instance, lose most of their nutritional and medicinal values by the time they have dried. For herbs such as these, and whenever practical, I like to make a fresh herb tincture right at the site of harvest. This ensures that most of the medicinal constituents and plant energies will be retained as I found them in nature. If you cannot make fresh herb preparations on the day of harvest, your herbs can be refrigerated for two or three days before processing. Be sure, though, that the herb is not wet when you stow it in the fridge, and (once again) don't store fresh herbs in anything plastic. Otherwise, you guessed it—refrigerator compost!

Gelatin Capsules

Dried herbs can be ground into a fine powder and put into gel caps. Though gel caps are handy for travel and convenient for other situations, herbs used in this manner have some drawbacks. First, you can't squeeze much herb material into a gel cap. Typically, the user will have to swallow from four to twelve gel caps per dose. If you are using herbs that are gentle and tonic in nature—such as dandelion, burdock, or alfalfa—for general health maintenance, you would have to fill a multitude of gel caps to make up a one-week supply. Second, gel caps must completely dissolve in the digestive tract before the herb itself is available to the body. After the gel cap has dissolved, the herb material must be digested and successfully absorbed *before* it reaches the colon. This usually means that the body uses only a small amount of the herb you have ingested. Studies indicate that for some people, gel caps may pass through the entire digestive tract without breaking down. A third problem with gel caps is that, contrary to what many people may think, the powdered herbs they contain have a limited shelf life. Whether gel caps are packaged in vacuum-sealed containers or not, oxidation is working inside each capsule. The herbs contained in many commercially available gel caps look pale or bleached because they contain herb dust.

Encapsulate your herbs if you must, but my advice is to make a tincture instead.

Alcohol-Based Tinctures

With few exceptions, an alcohol tincture is by far the most versatile, potent, and readily usable form of herb preparation. Alcohol tinctures offer the advantages of maximum potency and unlimited shelf life. Provided they are stored in airtight, glass containers, away from light, a properly made herb tincture can easily outlive its user.

The concentrated, liquid nature of tincture preparations allows for quick and complete absorption into the body. Kept in 1-ounce dropper bottles, tinctures can be carried anywhere and are as easy to use as squirting the appropriate dosage into some water, juice, or tea, or directly into your mouth.

Besides internal applications, alcohol tinctures can be used externally as liniments. They can be diluted into water and used as an eye, hair, or skin rinse, or added to olive oil and beeswax to make salves and ointments.

A tincture is prepared by soaking a measured quantity of fresh or dried herb in a certain proportion of alcohol and water, a combination called a "menstruum." The alcohol is a solvent, which breaks down the plant material and releases the active constituents into a liquid base, called a "tincture."

Two basic methods are used for making tinctures: percolation and maceration. For percolation you use a specially designed, cone-shaped glass vessel. Place the herb material in the cone, then pour the menstruum over it, and let the mixture stand for a short time until the liquid trickles (percolates) through the plant material and out of a tube at the bottom end of the vessel.

Making herb tinctures by maceration takes more time, but is much simpler than percolation and requires no special equipment. To make a macerated tincture, place herb material in a glass, plastic, or other nonmetallic container, pour the menstruum over the herb, cover the entire mixture with an airtight lid, and allow it to soak for two weeks. Then strain the menstruum from the plant material (called the "marc") to produce the finished, liquid product.

The proportions of alcohol and water in making an herb tincture vary according to the solubility of the herb. Some herbs, such as milk thistle seed, require a high percentage of alcohol to extract optimum quantities of active constituents from their cell structures. Others, such as raspberry leaf, require little or no alcohol to extract their useful compounds. Determining what percentage of alcohol you need is easy, because these determinations have already been done for you. Most of the formulas used in contemporary herbal practice are from the *United States Pharmacopoeia* and other standard references that have been used in the pharmaceutical production of simple herb extracts for most of the twentieth century. Herbalists and laboratories throughout the world still use this information, and it can be found in several books about herbs and herbalism.

In the "Care after Gathering" section for each plant described in this book, you will find the percentage of alcohol necessary to make a good tincture from the featured herb, with the appropriate ratio for herb material versus liquid menstruum. The monograph for arrowleaf balsamroot, for example, specifies that to make a tincture, you will need a menstruum comprised of at least 70 percent alcohol. That is the strength of menstruum you will need to extract the medicinally active components of the plant. You will also read that you will need to combine the chopped root and the menstruum at a ratio of 1:2 if the roots you are using are fresh, or at a ratio of 1:5 if the roots have been previously dried. This means that to make a tincture from the fresh root of arrowleaf balsamroot, you will need to proportion one part chopped herb material (dry weight) to two parts of menstruum (by liquid volume). For instance, 16 ounces of root must be covered with 32 fluid ounces of menstruum to make a tincture (and the menstruum must be 70 percent alcohol). For dried root, with the ratio 1:5, 16 ounces of dried root material will require 80 fluid ounces of menstruum to make a tincture.

How do you make a menstruum?

The most accurate and inexpensive way to make a menstruum is to buy pure grain alcohol (brand-name "Everclear") at your local liquor store. Everclear is 95 percent pure grain alcohol (190 proof). For simplicity, let's just call it a pure, 100 percent alcohol. To make a 70 percent alcohol menstruum, you would add 30 percent to its volume. If you need a 50 percent menstruum, add an equal amount of water. It's that simple. Everclear is not available in some parts of the country, so the medicine-maker might have to shop around. For tinctures that require 70 percent alcohol, uncut 151 proof rum will work fine—it is 75 percent alcohol. For a 50 percent menstruum, you can use 100 proof vodka—it is 50 percent alcohol—and so forth.

Here's a highly simplified version of making a tincture. Chop or grind some fresh or dried herb as fine as possible, and then pour enough menstruum (of the appropriate strength) over it to just barely cover it. Cover the container, let it sit for two weeks, strain it through a lint-free cloth, and store it in an airtight (preferably amber colored) glass jar—you now have an herb tincture. The primary rule in this simple method of tincture-making is always to use good-quality herbs, the correct percentage of alcohol in your menstruum, and to make sure that all of the herb material is completely covered with the menstruum before you let it stand for two weeks. Meet these requirements and your end product will be a medicinally active tincture.

Glycerin-Based Herb Extracts (Glycerites)

Glycerites are made by the same methods and mixture ratios as alcohol-based tinctures, but with vegetable glycerin in place of alcohol. To make a glycerite from dried herbs, dilute the syruplike glycerin 40 to 50 percent with water. To make a glycerite from fresh herbs use pure glycerin as the menstruum, as the glycerin will draw water out of the plant material to thin the finished tincture.

The advantage of glycerites over tinctures is their palatability—glycerin tastes sweet, with a flavor and consistency similar to light corn syrup. The big disadvantages with glycerites is their limited shelf life (one to two years), and in many cases, their strength. Glycerin is not as strong a solvent as alcohol, and many herbs will not take well to a glycerin menstruum, particularly if they are of a resinous nature. However, some herbs take very well to glycerin, especially those that are highly water-soluble. And even in many cases where glycerites are relatively weak, their lack of potency can be countered with increased dosage. The plant profiles in this book note if a plant does not take well to glycerine. You may assume that if a description lacks such a note, the plant is suited to a glycerite.

Vegetable glycerin is refined from coconut oils, and despite its syrupy-sweet flavor, it is metabolized not like a simple sugar but in a manner similar to a triglyceride. Glycerites can be used safely and easily for children, animals, people with alcohol intolerance, and people with diabetes. Vegetable-based glycerin is available through most herb retailers and at some drugstores.

Water Infusions (Herbal Teas) and Decoctions

Infusions are made by steeping plant material in hot water. The disadvantages of teas lie in their strength, sometimes terrible flavor, and the inconvenience of drinking them two or three times per day. A strong cup of hot tea makes good sense when the ingestion of hot liquid adds to therapeutic goals. For a chest cold associated with a buildup of thick, hard-to-move bronchial gunk, I have found that a steaming cup of yarrow, ginger (wild or culinary), and mullein leaf tea really helps to move stuff out because of the formula's expectorant (coughing that is productive) and diaphoretic (sweat-inducing) actions.

Water solubility defines whether an herb can be used in an infusion. Some herbs just won't take to hot water. When something hot and steamy is indicated and the herbs you need will not take to tea in either fresh or dried form, a teaspoon of so of herb tincture can be added to the hot water to serve the same purpose. Or, the herb can be "decocted" instead.

A decoction is an infusion that requires gentle simmering to make a strong enough preparation for therapeutic use. This is required of many roots when the plant material is too insoluble in water to allow a strong enough infusion. Prepare decoctions using a minimum of heat—simmer the herb for about 15 minutes at a temperature just below the boiling point. Too much heat will destroy many of the plant constituents you're after. Decoctions are used in the same manner as water infusions, but they are much stronger.

Oil Infusions

Produce an oil infusion by completely covering chopped herb material with olive oil and allowing it to steep in a covered, nonmetallic container in a warm (60 to 80 degrees Fahrenheit) location for at least one month. Then press out the oil and store it in the refrigerator for up to one year; a small amount (2 to 5 percent) of vitamin E oil can be added as a preservative. Although you can use several choices of vegetable oil, olive oil has its own preservative qualities, is relatively affordable, and is nourishing to the body both internally and externally.

Oil infusions are especially useful for topical applications, as they soothe and protect the affected area while holding herb constituents at the site where they are applied. Use extra caution when applying oil infusions to burns or infected areas, as they may seal in heat and bacteria. Make sure that any possibility of infection has been addressed with a good, clean antimicrobial preparation, and that a burn has cooled entirely before applying anything to it.

Poultices

A poultice is made by mashing plant material, usually dried foliage, with enough water or vegetable oil to make a wet, pesto-like paste. Poultices are especially good in topical field applications, when other preparations are impossible. Poultices cool and soothe the areas where they are applied.

Salves and Ointments

Salves and ointments are thickened oil infusions. To make a salve, use beeswax as the thickening agent. Thinner ointment preparations typically use coconut butter, particularly if they are for suppository applications. Gently heat the oil infusion (a glass double boiler is ideal), and melt the beeswax or coconut butter into the liquid until the cooled product is the desired consistency. For salves, a ratio of 1 ounce of beeswax or coconut butter to 8 ounces of oil usually makes a good starting consistency. If the salve or ointment is too thick or thin when it has cooled, heat it again and adjust the quantity of beeswax or coconut butter. Suppository molds and salve tins are available through herb catalogs and at some drugstores.

As with oil infusions, use extra caution when applying salves or ointments to areas of the body that are burned or infected, as they may seal in heat and bacteria. Treat the infection first with a good, clean antimicrobial preparation, and make sure that a burn has cooled entirely before you apply anything.

Fomentations

A fomentation is used for topical applications when a water or oil infusion is to remain on a specific body site for a specified time. Place gauze or other cloth material over the area, and pour the infusion onto the dressing until it is soaked. Mustard plasters, castor oil packs, and other traditional remedies follow this method.

GUIDE TO MEDICINAL PLANTS

∽

Thousands of useful plants grow in North America; selecting a few to discuss in this book was not easy. I based my choices on their plant chemistries and on valuable insights the plants offer into the environment we share with them.

About 75 percent of the plants described in this book can be found in their respective ecological niches throughout the United States and southern Canada. Many species discussed here can be cultivated outside their native environments, and some have escaped cultivation in areas where they do not occur naturally.

Plant Identification

This book lays a solid foundation for plant identification that you can use as a basis for further studies. Read the text, look at the photographs, experience the plants in the wild, and consult your botanical key. The glossary and bibliography will assist your continuing studies about medicinal plants.

A botanical plant key is the safest and most accurate way to identify a plant in the wild. Scientific terminology and practical complexity of botanical keys, though, can make them difficult to use and intimidating for many people. But if you plan to harvest and ingest plants, get a botanical key for your bioregion, and learn how to use it.

A botanical plant key takes its user through a progressive process of elimination. At the front of the key, plants are divided by the general characteristics of their botanical order. After identifying these characteristics, the user is offered one or more choices of physical characteristics that will place the plant into a family. Next, more choices are offered as clues of which genus the plant is a member. After you identify the genus, the real hair-pulling begins—singling out the minute features of the plant that will define its species. The problem with keys isn't in the process of identification, but in the reader's required learning of Latin, Greek, and some really tongue-twisting scientific jargon. If you lack a college-level botany course, try this: find a common plant that an expert has positively identified (perhaps a catnip plant at the local nursery). Look up the scientific name for the plant in your botanical key, and work backwards from the characteristics that match the key. What you will likely find is that the primary identifying features of the plant you are interested in will begin to jump out at you. For instance, whenever I see four-sided stems combined with opposite leaves and flowers in the leaf axis or in terminal spikes, I turn directly to the Mint family pages of my plant key because those are the distinguishing characteristics of the mint family that I easily recognize.

How to Use This Book

Headings
Each plant profile begins with headings giving the plant's most widespread common name (catnip), Latin genus and species *(Nepta cataria),* common family name (Mint family), and Latin family name (Labiatae).

I have chosen to list the plants in this book alphabetically by their common names. Although unconventional in scientific books, the simple and recognizable format of this book will appeal to people with varied levels of plant familiarity. If you want to find a plant by its family name or Latin name, refer to the index.

Common names tend to be generic, with the same name typically referring to dozens of unrelated plants. As you begin to recognize the plants by their common names, also learn their Latin names—it's not hard. Start by learning the genus name of each plant. The genus refers to a group of species within a plant family. In many cases that is all you will need to know; learning the exact species (specific member of a genus) can come later.

You may notice that some plant profiles list only the name of the genus. In these cases, either the entire genus or several of its members are medicinally useful and several such species will be discussed in the description.

Icons
Icons that appear to the left of the heading for some species convey important information at a glance.

 The United Plant Savers (UpS) logo identifies species that UpS currently considers most sensitive to human activities. Please help preserve these species.

 The herbicide-watch icon indicates plants that, because of their weedy nature or their association with weedy plants, may carry herbicide residue. Carefully examine the health of these stands before you harvest anything.

Other Names
Because most plants have many common aliases, I have included some other well-known common names under this subheading.

Parts Used
The parts of the plant that are used for medicinal purposes.

Actions

The term *actions* relates to the interactions a plant has with the human body—for example, an analgesic relieves pain; an anti-inflammatory reduces inflammation. This section identifies what I believe are the most important medicinal attributes the plants have to offer. It is not a complete overview of their therapeutic possibilities. The glossary provides definitions of medical terms that you may find unfamiliar.

Description

A basic physical description of each plant, written in the simplest terms possible, emphasizes distinguishing characteristics. Some plant characteristics are difficult to convey with anything but botanical terminology. When you encounter an unfamiliar botanical term, please refer to the glossary. The labeled drawing on page 234 illustrates the anatomy of a flower.

Habitat and Range

This section provides loose parameters for each plant's habitat and range. I refer to general ranges because the range of many plants changes constantly with alterations in weather patterns, seed transport by migrating animals, introduction by humans, and countless other influences. Many plants are adaptable to various habitats and mutate or hybridize in their efforts to survive in a changing world.

Applications

I briefly discuss some of the most effective and earth-conscious uses for each plant. This information is based on my experience and opinions as well as on scientific data, anecdotal evidence, traditional beliefs, and other data that I find interesting and pertinent to this book. I have written it from my own perspective, that of an herbalist who uses herbal medicines in my own health maintenance.

Besides listing what I view as the plant's best medicinal attributes, this section also provides information about known side effects and common misuses, and some food for thought about therapeutic possibilities that are still being investigated. As you read this information, bear in mind that no two human bodies are alike. An herb that brings favorable results to one person may cause adverse reactions in another. Our knowledge of medicinal plants and the chemical compounds they contain is constantly evolving, as are the plants. Plants that may have benefited a Celtic warrior may not necessarily work for a computer analyst.

For this reason, I have not included doses in the herb profiles of this book. Determinations of dose, dose frequency, and duration of therapy are based on the situation at hand and the specific needs and body type of the individual. Many herb books provide the reader with general dose parameters, but such parameters only serve as a vague idea of how much of an herb might be needed. Most herbal pharmacopoeias recommend generalized doses of between 2 and 6 milliliters of an herbal tincture, two or three times daily for a 150-pound adult. For children, the dose is usually prorated to the subject's body weight. For instance, a 75-pound child would require half as much as a 150-pound adult. To make an herbal tea, steep 1 to 3 teaspoons of dried herb in 8 ounces of hot water. Prepare and ingest the tea two or three times daily. However, exact dosage can only be determined by a trained and experienced professional after a thorough assessment of the individual and cannot be contained in the pages of a single book.

I encourage you to further your knowledge by reading some of the excellent texts recommended in the bibliography. If you choose to use plants in any nutritional or therapeutic capacity, exercise a liberal dose of caution and common sense. Take whatever you read in *any* book with a grain of salt until you understand how the plants you are studying will behave in or on your body. I am not a licensed physician or health care practitioner. If you have a health problem or wish to take preventative steps toward your future well-being, consult a licensed health care professional.

Alternatives and Adjuncts

In this book, an "alternative herb" means one that will usually serve as an adequate substitute for another. Arrowleaf balsamroot, for example, is often—but not always—an effective substitute for echinacea.

An "herbal adjunct" is an herb that will contribute to the overall effectiveness of the primary herb. Marshmallow, for instance, will usually soothe an irritated throat when used with an expectorant, such as arrowleaf balsamroot.

In this section I introduce a few of the many herbal alternatives for each plant described, as well as some herbs that I believe work particularly well with the primary plant. This list is not exhaustive, because there are no definitive rules for combining herbs; no two sets of circumstances are alike. You will need to consult other books to learn about some of the alternative and adjunct plants named here.

Again, I strongly urge you to get to know your bioregion. Many of the plants in this book are abundant in some areas but scarce in others. Learning about which alternative plants grow in your area, can be grown in the garden, or can be obtained through cultivated sources is paramount to the earth-conscious use of herbs.

Propagation and Growth Characteristics

This section draws on my research and experience with the cultivation and natural reproduction of each plant. It is intended to assist you in propagating plants you harvest, following nature's way as closely as possible. This section underscores the delicate natural balances needed to ensure a plant's survival.

I believe in growing large herb gardens. This section provides the basic information you need to make your herb garden a success.

Gathering Season and General Guidelines

Earth-conscious guidelines and suggestions about how, when, and if you should gather plants from the wild include practical information based on my experience with each plant. No two hillsides, meadows, or hollows are alike, and low-impact harvest techniques that work at one site might result in environmental crisis somewhere else. The key to ethical wildcrafting rests not in adhering to a fixed set of standards but in working carefully within the natural rhythms and balances of each bioregion.

I hope the information in this section—gleaned from my trial-and-error experience in the field—will help you avoid mistakes that could harm our plant and animal friends.

Care after Gathering

This section outlines preparation and storage methods that will assure optimum potency, shelf life, and the overall usefulness of the herbs you gather. The techniques and formulas I have provided for making tinctures, oils, infusions, and so on are universal throughout the Western herbalist community. Most of them are based on pharmaceutical standards of fifty or more years ago, when herbs predominated in the American pharmacy. Please consult other references to identify formulas and techniques that may be specific to your interests.

Plant-Animal Interdependence

Sustainable use of wild plants requires us to grasp the concept that all organisms, from amoebas to moose, in some capacity depend upon one another. This section explores these relationships between the featured plant and other plants and animals.

The survival of the collective body we call an ecosystem depends on an endless diversity of natural checks and balances. Each plant, animal, insect, mineral, and microorganism is a critical piece in a complex puzzle, and if one piece is missing, the entire puzzle may be at risk of falling apart.

Tread Lightly

In this section I talk about human impact considerations that I feel are of special significance to the healthy survival of the plant.

Warnings

At the end of some plant profiles, I have noted any special considerations about using the plant safely. Profiles that do not include a warning subheading are generally regarded as safe, but remember that anything is potentially toxic if used in excess, in the wrong context, in conjunction with certain drugs, or by people with hypersensitivity to one or more compounds in the plant. The information under this subheading is not comprehensive enough to determine whether an herb will be appropriate for each reader. If you wish to use any of the herbs in this book, seek proper training and supervision from a qualified clinical practitioner.

Alumroot
Heuchera species

Other Names: Mountain saxifrage, alpine heuchera, poker alumroot
Parts Used: The root
Actions: Astringent

A perennial with a short bloom duration, this plant typically grows as a cluster of ovate to heart-shaped, sometimes maplelike, basal leaves protruding from steep banks or rocky areas. The leaves are similar to currant leaves and generally remain on the smaller side of their ½- to 3-inch range. Alumroot plants are usually small—6 to 12 inches high when mature.

Alumroot flowers are small and cup-shaped, with five simple petals ranging from greenish white to light pink. The leafless, erect flower stalks reach much higher than the leaf cluster, sometimes as tall as 20 inches, each bearing one to several flowers that grow along only one side of the stalk. After a brief bloom period, the flower stalk looks like a brown or rusty twig, a characteristic that helps with identification in late fall and winter.

The root is proportionately large and covered with scaly brown, dead plant material. The inner pith of the root is pale and has a strong astringent-acidic flavor because of its extremely high tannin content.

Habitat and Range: Alumroot, a cliff-clinging little plant, is common on steep, rocky hillsides and forested areas where at least 50 percent is shaded and moist most of the time. It needs rich organic matter at the roots for support in its often harsh environment, and rocks and other structures for a foothold. These specialized environmental requirements help define this plant: it often grows from niches and fissures where lichen by-products and other debris have sufficiently accumulated from water runoff.

In the northern Rockies, alumroot grows at elevations as low as 3,500 feet, but it usually grows in more moderate climates, and much higher, sometimes up to 9,000 feet. Wet, shady, rocky banks and slopes are the first places to look for this plant.

Applications: One of nature's heavyweight astringents, alumroot may contain up to 20 percent tannins. To call it a strong astringent is an understatement. It is so powerful at shrinking tissues that it would be my herb of choice if I ever took up head shrinking. Even a small piece of the root chewed or placed on the tongue brings on an instantaneous case of herbally induced cotton mouth. This action, whether pronounced, as with alumroot, or relatively subtle, as with cranberry juice, reveals much about the healing characteristics of tannin-rich plants. Tannin-bearing herbs are useful both internally and externally to stop bleeding or to prohibit the transport of fluids across mucous membrane barriers. The stronger the tannin concentration, the faster and more dramatic this action.

Alumroot   *Heuchera* species

For external use, finely grind the dried root with a mortar and pestle or an electric coffee grinder for use as a styptic powder to pinch off bleeding of minor wounds. This powder is also useful for relieving surface inflammation from insect bites and stings, minor burns, or acute cases of dermatitis (such as poison ivy). The powdered root will treat swollen hemorrhoid tissues, either wet and applied as a poultice or infused and applied with gauze. Externally, it is safe to use wherever a strong astringent is indicated.

Herbalists sometimes use alumroot to check minor internal bleeding (a condition requiring the attention of a trained practitioner) or as a symptomatic remedy for diarrhea. Alumroot tightens the mucosa of the intestine, inhibiting the entry of fluids into the lower gastrointestinal tract. As a mouthwash or gargle, alumroot tea (made by steeping 1 teaspoon of fresh or dried, chopped root in 8 ounces of hot water) will relieve swollen gums or a sore, irritated throat.

Potential problems with using alumroot internally are the same for any herb high in tannins. Too much for too long can irritate the digestive and urinary tracts, including the kidneys. Most herbalists agree that internal use of this herb should be reserved for short-term applications in individuals who are not sensitive to astringent substances, and only for cases where a weaker alternative is not available or is not strong enough to do the job. Alumroot should not be ingested during intestinal constipation or in cases of pre-existing kidney disease or chronic inflammation of the urinary or digestive tracts.

Alternatives and Adjuncts: To stop internal or external bleeding, yarrow is my herb of choice. For insect bites, burns, and internal use against diarrhea, consider using plantain, pyrola, rose bark, storksbill geranium, or uva-ursi. For internal applications, herbalists often combine alumroot with demulcent herbs, such as plantain, marshmallow, or fireweed. The mucilaginous nature of these plants will lend a soothing, lubricating, and protecting action that will cut alumroot's harsh, potentially irritating astringency.

Propagation and Growth Characteristics: This plant's habitat allows it to reseed only by coincidence. The seeds are tiny and require light to break dormancy. If the seeds do not find their way into the same nook as the parent plant, gravity or runoff usually carry them downslope. The seeds must then settle in a spot with proper soil structure where they can receive the correct amounts of light and water to germinate and take root—slim odds for a seed smaller than a pinhead!

According to Janice J. Schofield, author of *Discovering Wild Plants,* the crown of the plant can be successfully transplanted.

Gathering Season and General Guidelines: Collect the root after the plant has gone to seed (June to September). Although late-season roots contain higher concentrations of active constituents (tannins), alumroot should be gathered as early as possible after it has dropped its seeds. This allows more time for the transplanted crowns to take root before winter. To ensure the highest degree of transplant success, replant the crowns exactly where they were.

Care after Gathering: Like all roots gathered from a damp environment, alumroot is susceptible to mold. Cut large roots in half lengthwise, then follow standard procedures for drying roots (see "Harvesting and Handling Herbs in the Field"). After the root has dried, it can be ground into powder and stored in an airtight, nonmetallic jar for future use. The dried, powdered root is an excellent styptic powder for shaving nicks, cuts, and scratches. The root, fresh or dry, can be tinctured for indefinite shelf life.

∾Tincture
Fresh root: 1:2 ratio in 50 percent alcohol and 10 percent glycerin.
Dried root: 1:5 ratio in 50 percent alcohol and 10 percent glycerin.

Plant-Animal Interdependence: Where alumroot grows from fissures in rocky cliffs, it creates a microcosm of minute organisms. These delicate, vertical minihabitats maintain life in an otherwise uninhabitable environment. The proportionately extensive roots of this plant serve a critical, sometimes exclusive, role in erosion control. The ecoherbalist needs to fully appreciate the fragile balances this plant supports.

Tread Lightly: Alumroot commonly lives in environments better suited to mountain goats than people. Protect habitat and avoid vertical adventures by gathering alumroot from areas with easy access and solid footing. If you gather roots from a steep slope, plan your approach carefully and maneuver slowly and cautiously to avoid unnecessary slipping and sliding. Moving a single rock or dislodging a tuft of moss could compromise next year's growth. Monitor the effects of your gathering and transplant efforts over time.

American Ginseng
Panax quinquefolius

Ginseng Family
Araliaceae

Other Names: Panax ginseng, wild ginseng, 'sang
Parts Used: The root; sometimes the leaves
Actions: Adaptogen, stimulant, antidepressant, tonic, and demulcent

Only nine species of *Panax* exist worldwide. The two North American species, American ginseng *(P. quinquefolius)* and dwarf ginseng *(P. trifolius)*, look similar and share the same habitat. These two perennials may reach 2 feet tall. In both species, the palmate leaves are divided into three to seven broadly lance-shaped segments, each with distinctively toothed margins. The whitish flowers are flattened umbels, formed shortly after the plant emerges in early spring. The flowers later develop into a cluster of bright red, two-seeded berries.

Habitat and Range: American ginseng's hardwood forest habitat is disappearing almost as fast as the plants themselves, to the buzz of the logger's chain saw and the grumble of bulldozers. This plant once flourished in North America, from the eastern provinces of Canada south along eastern New England into the Appalachian Mountains, Georgia, and Oklahoma. Today, the healthiest stands of American ginseng are in Minnesota, parts of the Carolinas, and the northern half of the Appalachians.

Applications: American ginseng is said to increase physical and mental performance, build disease resistance, raise low blood pressure, increase available energy, and help the body adapt to various forms of physical or nonphysical stress. Given such broad parameters of traditional use, it's no wonder that this plant is one of the most sought after native American plants. The idea that any plant could do all that is asked of ginseng is unreasonable and leads to waste.

From a holistic perspective, effective treatment of a person's general debility or chronically depressed immune system begins with a tedious assessment of his or her mind, body, spirit, emotional state, life and genetic histories, and environment. The next step may require adjustments in nutrition, an increase in daily exercise, changes in habits, and countless other variables. After identifying the root cause of the problem, an herbalist can select plants to satisfy certain requirements lacking from the holistic picture. Taking a dose of ginseng every day because you feel run-down after lunch is counterproductive to the herbalist's goal of living in a state of healthy balance. When used properly, ginseng is an extremely useful medicinal device. In my experience and opinion, ginseng is best used where it accentuates the actions of other herbs. Used with echinacea to ward off a viral infection, for example, ginseng compliments the immune-supporting activities of the echinacea by increasing echinacea's metabolic responsiveness.

The demand for wildcrafted roots, which some people believe are superior to cultivated ones, will continue to drive the market for American ginseng. The argument over whether wild ginseng is better than its cultivated counterpart is overshadowed by the serious, urgent reality that if we don't stop supporting the use of wild American ginseng, it will soon be gone. This plant represents a major part of our herbal heritage—it's time for us to join in healing what has healed us for generations.

Alternatives and Adjuncts: If you use ginseng, buy it from a cultivated source. If you are looking for an immune stimulant, consider using echinacea, astragalus, arrowleaf balsamroot, or shitake mushrooms. Astragalus *(A. membranaceous)* and Siberian ginseng *(Eleutherococcus senticosis)* are also "adaptogenic" herbs and are said to increase vitality much the way American ginseng does.

Propagation and Growth Characteristics: Successful cultivation of ginseng can be tedious and very expensive (as much as $30,000 per acre!). It is

American Ginseng   *Panax quinquefolius*

The fall fruits of American ginseng. As the fruits ripen, the small terminal leaves eventually die back.
—Paul Strauss photo

not impossible, however, especially if you have an ideal piece of natural habitat. The best place to grow ginseng is on a north- or east-facing hillside, in a healthy mixed hardwood forest that offers deep, humus-rich soil, mostly shaded conditions, and natural mulch. Ideally, soil pH should be in the range of 5.5 to 6.0. Cultivate the plant by root cuttings or from seed. Because the plants will not reach peak medicinal potency for at least six years after they sprout—which might take two years—I would opt for planting roots that are at least three years old. Growing "woods-cultivated" ginseng can be fun and rewarding, both financially and spiritually. Before you invest in root stock, though, learn the "ins and outs" from one of the experts listed in the "Resource Guide" of this book.

Gathering Season and General Guidelines: Ethical wildcrafting of wild American ginseng is no longer possible except plants rescued from areas where development will wipe them out anyway. Every effort to transplant the roots into a viable habitat should be exhausted before using them as medicine. Buy from cultivated sources, or better yet, grow your own and help to assure a future for this plant.

Harvesting ginseng you have nurtured for several years is a thought-provoking experience. You have watched it push through the earth and compete with the elements year after year. Then, when the time comes, you face the sobering decision of whether to let the parent plant live (some plants may survive more than fifty years) or to sacrifice it. If you decide your needs outweigh the sacrifice, proceed with the harvest: dig the root with care and a prayer, wash it with a mist of clean water, and let the root, in whole form, dry slowly—for at least six weeks. When thoroughly dried, the root is rock hard and looks like a tiny, withered old wise man.

Care after Gathering: Properly dried ginseng roots will keep for several years if stored in a paper bag. Do not store them in plastic or they could develop mold. Leave the roots in whole form until you are ready to use them, to assure maximum shelf life. When a need arises, the root can be ground, hammered, or shaved for use in teas or other formulations.

Plant-Animal Interdependence: Any plant that survives in the forest as long as American ginseng surely plays a critical role in the ecosystem. Many birds and other animals relish the leaves and berries of American ginseng, which will test your natural respect and personal conviction if you have nurtured the plants from seed!

Tread Lightly: At the rate this plant and its habitat are vanishing, it will take the hearts and hard work of many caring people to bring the population back. If you want to be part of this effort, please join United Plant Savers. See the "Resource Guide" for more information.

WARNING: Ginseng may elevate blood pressure.

Angelica
Angelica species

Parsley Family
Umbelliferae

Other Names: Kneeling angelica, osha del campo
Parts Used: The root
Actions: Antispasmodic, carminative, expectorant, diaphoretic, and diuretic

A hollow-stemmed plant, angelica stands 2 to 5 feet tall when mature. The large leaves are divided into pairs of lance-shaped leaflets, each about 2 inches long with serrated edges and smooth or lightly hairy surfaces. The leaf veins of angelica terminate at the outer tips of the leaf-edge serrations, unlike those of the highly poisonous water hemlock (*Cicuta* species), which has leaf veins terminating at the *bottom* of the leaf serrations (see photos). The stems of angelica are usually, but not always, purple or reddish at the base. Do not use stem color to positively identify angelica.

The aroma of this plant is similar to garden variety lovage, but with a conifer overtone. It is a scent that becomes distinctive as you gain familiarity with the plant and is useful in distinguishing it from poisonous look-alikes.

Like other members of the Umbelliferae family, the flowers of angelica are umbel shaped and look like a small burst of fireworks. The umbels consist of hundreds of tiny, white flowers, which later develop into white to very light brown, double-sided seeds reminiscent of angel wings, a characteristic that easily distinguishes it from water hemlock (water hemlock has small, corky, somewhat kidney-shaped seeds obviously different from angelica seeds).

The root of angelica is large and fleshy, medium brown, and distinctively tapered (like a carrot). The inside of the root is usually solid. Cut the root in half lengthwise to see if the interior is solid; if the interior is chambered, you may have mistakenly gathered the root of water hemlock. Despite what some books say, this is not a conclusive way to distinguish angelica from water hemlock—either species may have solid or chambered roots.

Examine the stand of plants carefully before gathering angelica. If water hemlock is present, don't gather angelica! Several angelica species grow throughout North America. All are useful—check your local references for the angelica in your area.

Habitat and Range: Angelica generally grows close to water but usually not standing in it. It prefers the fringes of wet, boggy areas and is common along stream banks, springs, and roadside ditches.

Angelica grows in various exposures, but most species prefer an equal mix of sun and shade. Angelica grows from coastal elevations to the upper reaches of the subalpine zone in the northern two-thirds of the United States and north into Canada and Alaska.

Applications: In Traditional Chinese Medicine, a preparation of *A. sinensis,* an Asian species, is made into a medicine called "Dong Qui." Dong Qui has been a fixture in Traditional Chinese Medicine for thousands of years. In China, therapeutic applications of this ancient remedy vary widely, but here in the West, it is most commonly used in the treatment of menstrual cramping and discomforts associated with menopause.

Whether our American species of angelica are useful as effective substitutes for the Asian *A. sinensis* is the subject of debate among Western herbalists and practitioners of TCM, but most will agree it is an excellent digestive system antispasmodic, useful for easing gastric bloat associated with generalized cramping, spastic colon, and related ailments. Many herbalists also recognize a strong antifungal activity with this herb, and sometimes use it to treat digestive candidiasis (in tincture or tea form). An infusion of the dried root can be used in a footbath to inhibit athlete's foot fungus, or it can be cooled and used in a vaginal douche.

A serious risk for self-reliant herbalists is that angelica can be very difficult to distinguish from its deadly relative, water hemlock. Obtain this plant

Water Hemlock   *Cicuta douglasii*   Angelica   *Angelica* species

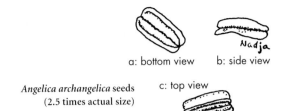

a: bottom view   b: side view

Angelica archangelica seeds
(2.5 times actual size)

c: top view

Angelica   *Angelica archangelica*

only from certified, organically grown sources, or wild-harvest it only if you are an experienced wildcrafter. One mistaken identity could be fatal.

Alternatives and Adjuncts: For antifungal applications, bee balm, western sweetroot *(Osmorrhiza occidentalis)*, or usnea lichen (*Usnea* species) are good choices. For abdominal antispasmodic applications, look to black cohosh, chamomile, peppermint, or cow parsnip seed.

Propagation and Growth Characteristics: Angelica is a perennial that may require two or more seasons to produce viable seeds. Alluvial action and animals distribute the seeds, which require cold stratification to germinate.

This plant likes rich, moist soils that range from slightly alkaline to slightly acidic, depending on the species. It is easy to grow in a garden, which I strongly recommend for positive identification of this plant.

Gathering Season and General Guidelines: Gather angelica as late in the season as possible—late summer to early fall, after the seeds have ripened and begun to fall off. Do not wait too long, as positive identification becomes difficult after the seeds are completely gone. Because the root is the main portion that is used, leave the upper parts to decompose where the plant was dug. Mature roots are often more than two years old, so use multiple gathering sites and monitor the effects of your presence. To determine root maturity, carefully remove soil from the base of the plant. If the taproot is small, move on and leave that plant to grow some more.

Care after Gathering: Cut the roots lengthwise and dry them in a well-ventilated, dark place. Fresh root is not recommended; it is strong and may cause stomach upset. The dried root can also be ground for use in teas or gel caps.

∾Tincture
Dried root: 1:5 ratio in 50 to 65 percent alcohol.

Plant-Animal Interdependence: Angelica provides habitat for ground-nesting birds and small mammals, and concealment for deer, elk, moose, and bear. The plant is an effective pollinator attractor and should be left undisturbed during its bloom period. The juicy, green foliage composts quickly after the plant dies and contributes substantially to the soil structure.

Tread Lightly: The moist, riparian soils where angelica grows compact easily and host a dense proliferation of other plant growth. Wading through thick stands of vegetation requires careful, sometimes meticulous maneuvering to avoid damaging foliage and root systems. If you use wild angelica, find a stand that is easily accessed with minimal soil compaction.

> **WARNING: Angelica is difficult to distinguish from water hemlock, unquestionably the most poisonous plant in North America. Positive identification is essential and must be established through leaf or seed examination.**

Arnica
Arnica species

Sunflower Family
Compositae

Other Names: Leopard's bane, wolf's bane
Parts Used: All aboveground parts of the flowering plant
Actions: Vasodilator, vulnerary, rubifacient, and anti-inflammatory

Arnica is one of the first flowers to emerge in early spring. The flowers in this large genus are commonly bright yellow, 2 to 3½ inches wide, and daisylike. Each plant has one to three flower heads when mature. Two to four pairs of 1½- to 5-inch-long petiolate, opposite leaves grow beneath the terminate flowers. The leaves range from lance-shaped to a nearly circular heart shape. Most species have toothed leaf margins, and some have leaves that are one to three times as wide as they are long. The stems and leaves are smooth to moderately hairy and are generally pungent, with a strong piney-sage odor.

Some arnica species, such as *A. parryi,* produce flowers that completely lack rays, which makes the plants look plucked. All species are small plants, seldom more than 20 inches tall, with the majority of species less than 10 inches high. Arnica has a short bloom period and may not bloom at all during drought conditions.

Arnica often grows in large, dense patches.

Habitat and Range: This mountain plant prefers partially shaded meadows, hillsides, and wooded areas from about 3,500 feet to well above timberline. It loves soils rich in organic matter, especially conifer debris, and grows from the thick mats of needles beneath pines, firs, and spruces. It acts as an earth-regenerator, sometimes in profuse abundance, appearing three or more years after an area has burned or the topsoil has been disturbed but not displaced by erosion.

Applications: Arnica is well known for its reliable and effective use in such closed tissue injuries as sprains, bruises, closed fractures, ligament injuries, and hyperextended muscles and tendons. Various arnica gels, liniments, and other topical preparations are commercially available, and athletes who push their bodies beyond their limits (competitive masochists!) commonly use them.

The actions of arnica derive primarily from volatile oils in the flowers and foliage. Those oils are readily absorbed through the skin and work quickly to dilate peripheral capillaries, allowing for increased circulation of lymph and blood in areas where injury is causing congestion of fluids. This, in turn, helps the body work toward a speedier recovery. People who use this herb for the first time in remedial treatment of closed injuries are often amazed at how quickly it reduces swelling and accelerates the overall healing process.

To use arnica, apply the bruised flowers and leaves, a tincture, or a water or oil infusion directly to the injury site. Repeat this two to three times a day until swelling subsides and remains in check. Arnica should not be used for extended periods of time, though, or a rash may develop. If that happens, stop using it! Most herbalists agree that arnica should not be applied for more than a few days, depending on the tolerances and sensitivities of the individual.

Although some advanced herbalists use arnica extracts internally, this can be dangerous without proper training and experience, as this herb may cause ulceration and bleeding in the digestive tract if ingested in anything but minute quantities. Never use arnica in open wounds, as its dramatic vasodilating actions will likely increase external bleeding. Arnica should not be used on people with hemophilia or other blood-clotting deficiencies.

Alternatives and Adjuncts: For first-aid treatment of closed tissue injuries, there is no substitute for arnica. Many herbs work well as complementary adjuncts to this herb, though, depending on the crisis at hand. Saint John's wort is an excellent addition for closed injuries associated with nerve damage. Willow bark, poplar buds, meadowsweet, or red osier dogwood can be used to help relieve pain at the site of injury. A horsetail poultice or tea may help regenerate damaged connective tissues.

Propagation and Growth Characteristics: Arnica is a perennial that blooms during its second year and every year thereafter. It commonly grows in dense colonies, where a single rhizome may extend beneath the surface of the forest duff for several feet, sending up a proliferation of offshoots that look like dozens of individual plants.

Arnica reseeds itself effectively after it matures. Seeds can be started in moist, light sand, or root cuttings can be transplanted directly into the garden. When introducing arnica into the herb garden, try to duplicate the specific habitat where each plant originally grew—and beware of its tendency to spread.

Gathering Season and General Guidelines: Although herbalists mainly use the flowers of arnica, the entire aboveground plant contains the constituents needed for medicinal use. The best time to collect arnica is during the first half of its bloom period, when the flowers are fully open but still in good condition (May to late July). Because this plant may take two years or more to bloom, nonflowering plants should be left to grow. Do not collect arnica during drought or dormant periods when plants do not bloom, as the maturity of this plant is difficult to determine during those periods.

When gathering arnica, grasp the plant at the base of its stem, right at ground level. Snap the upper plant off, leaving the rhizomes behind for next year's growth. Wear gloves, as the volatile oils of this herb can be absorbed through the skin.

Care after Gathering: After gathering arnica during its bloom period, you may wonder what the purpose is in picking the flowers. Regardless of how tight the buds may be, you probably will return home with a bag of white and

Arnica   *Arnica cordifolia*

Arnica   *Arnica cordifolia*

yellow fluff. This appears to be a survival mechanism, as the fluff contains the viable seed, regardless of the cyclic interruption.

This herb is best used fresh. Arnica oil can be made in the field by cutting up the herb and placing it in a jar as you harvest, then covering the herb with olive oil. The oil should be covered and kept in a dark place for one month, then strained through a clean cloth and stored in a clean jar.

The plant can be tinctured fresh or dried (fresh is best), for topical use only.

∿Tincture

Fresh herb: 1:2 ratio in 70 percent alcohol. *Dried herb:* 1:5 ratio in 70 percent alcohol.

Plant-Animal Interdependence: Arnica is an effective and important pollinator attractor. It is host to a wide variety of insects that play integral roles within the ecosystem. The plant's extensive rhizomes aerate the forest floor and provide habitat for subterranean insects, rodents, and other creatures. In some areas, arnica is an earth regenerator; its extensive rhizomes hold the soil in areas where water runoff would otherwise compromise soil structure.

Tread Lightly: In areas where arnica grows through thick mats of forest debris, soil compaction can be a problem. Do not walk in these areas if it has been raining there. High-compost soils lose resiliency when wet, so arnica should be harvested only during dry weather, when the forest floor feels springy underfoot.

Because of arnica's sprawling, rhizomatous root system, wildcrafters need to be extra careful not to damage the root systems when walking among the plants. What may look like many arnica plants may be the offshoots of only a few. Visualize the root system of arnica as a living, subterranean net. Then do your best to avoid damaging the shallow roots by working only from the periphery of the stand.

WARNING! This herb may cause internal bleeding if ingested.

Arrowleaf Balsamroot

Sunflower Family

Balsamorrhiza sagittata

Compositae

Other Names: Balsamroot
Parts Used: The root
Actions: Antimicrobial, expectorant, disinfectant, and immunostimulant

Arrowleaf balsamroot has large, 3- to 5-inch-wide, bright yellow flowers that unmistakably remind us it is a sunflower. Large communities of balsamroot often bloom simultaneously, illuminating an otherwise drab hillside or grassland with a brilliant show of yellow. Flowers form on individual, unbranched stalks that grow above the basal cluster of leaves. The stalks have only a few small leaves, a characteristic that helps differentiate this plant from the similar-looking *Wyethia* species (commonly called "mules ears"), another yellow or white-flowered sunflower that shares habitat with balsamroot.

The leaves are large, often a foot long, and are distinctively arrow-shaped. The plant is covered with velvety hairs, giving the leaves a silvery gray

Arrowleaf Balsamroot *Balsamorrhiza sagittata*

appearance. The roots, thick and tapered, are up to 3 feet long, and may weigh 3 pounds or more when mature. The roots are resinous and have a rich piney aroma. Balsamroot blooms in early to midspring and typically is the first yellow flower to appear in its habitat. It has a long bloom period and may be the last yellow flower in late summer.

Habitat and Range: Balsamroot flourishes on dry, open hillsides with southern to western sun exposure. It is well adapted to shallow, rocky soils low in organic matter, and often stands alone on otherwise blank hillsides.

Various species of *Balsamorrhiza* are common to the coastal foothills and in mountain areas from California east to the eastern slopes of the Rocky Mountains and north into Canada, up to about 7,000 feet.

Applications: I regard balsamroot as a bioregional alternative to echinacea. Although the root of this plant is not as strong as echinacea in its immune-modulating activity, many herbalists find that it is very effective at boosting the frontline defenses of the body's immune system. This action may come from an array of polysaccharide constituents that act similar to those contained in the wildly popular echinacea. Balsamroot also lends some medicinal attributes that echinacea cannot. When harvested in early spring, the root of balsamroot contains copious quantities of thick, gooey, blue jean-ruining resins that have excellent expectorant and antimicrobial properties. Herbalists maintain that in fall, when the leaves die back and the energies and chemistry of the plant concentrate in the root, the plant gains its immune-stimulating and antiviral nature. Therefore, unlike echinacea, balsamroot can be harvested both in spring and fall, tinctured, then combined to make an expectorant-antimicrobial–immune modulating formula that will serve a much broader spectrum of needs than echinacea.

Alternatives and Adjuncts: Balsamroot combines well with coltsfoot, mullein, or grindelia *(Grindelia squarrosa)* for an expectorant. Marshmallow, fireweed, or false Solomon's seal can be added as a soothing adjunct for a dry, irritated throat.

Echinacea can be used as a stronger immune modulator. Bee balm serves as an antibacterial adjunct when upper respiratory irritations and coughs call for a strong, antimicrobial double whammy.

Propagation and Growth Characteristics: A perennial, balsamroot is hardy, drought tolerant, and slow growing. Though it reseeds itself readily, a large mature root can be more than twenty years old. Transplanting whole roots is usually impossible with older plants because they are too long to harvest without damaging them. The root crowns can be successfully transplanted. After digging the root, pull the crowns laterally away from the root as carefully as possible and transplant into the same area where you gathered it, or introduce it into the herb garden. Animals or spring runoff naturally distribute the seeds, which require stratification to germinate.

Gathering Season and General Guidelines: Dig the spring roots before the plant blooms. Fall roots should be gathered anytime after the plant blooms and has gone to seed, usually late July to late August. If you wait too long, the roots become dry and woody, making processing difficult and reducing the quality of the finished product.

Care after Gathering: Because this plant is a slow grower and the mature roots are generally large, plan on tincturing the fresh root for long-term storage. One root can easily supply a household with tincture for several years, and by making a tincture right away, you eliminate any risk of spoilage or loss of constituents through drying. Cut the roots on a board that you plan to use exclusively for herbs, because the sticky resins will get all over it. If you must dry the roots, cut them lengthwise two or three times and stack them loosely in a paper bag. The root may be ground for use in teas or decoctions. The dried root will keep for six to twelve months if stored in airtight glass jars, away from sunlight.

∾Tincture

Fresh root: cut into small chunks (1 inch or less); 1:2 ratio in 70 percent alcohol. Arrowleaf balsamroot does not take to glycerine. *Dried root:* 1:5 ratio in 70 percent alcohol.

Plant-Animal Interdependence: Where balsamroot grows on steep, sparsely vegetated hillsides, it plays an important role in protecting against erosion. The large roots hold shallow soil in place and support grasses and other important wildlife forage. The basal leaves die back in fall and provide precious compost to otherwise depleted soils.

The large, heat-resistant leaves provide food and shelter for a multitude of insects and small animals; deer and elk, too, may browse balsamroot. I have seen evidence of bears digging the roots soon after their winter naps. The large, yellow flowers are effective pollinator attractors, and help draw attention to less conspicuous plant neighbors.

Tread Lightly: The ecoherbalist must remember that balsamroot is a very slow grower and be aware of the long-term absence of the plant after harvest. Before gathering the plant, stop and think about the aftereffects—erosion, loss of forage, and reduced small-animal shelter, especially where balsamroot is the primary plant in its habitat. Hillsides should be monitored closely, particularly during the spring melt and rainy seasons. In areas where erosion is already a problem or animal use is high, balsamroot should be left to do its job of hillside maintenance. Those areas include places where livestock graze and where large herds of deer or elk feed on a limited range because of urban pressures. When possible, gather from a healthy stand of plants on relatively level terrain.

Bee Balm
Monarda fistulosa

Mint Family
Labiatae

> **Other Names:** Wild bergamot, sweetleaf, purple bee balm, horse-mint, wild oregano, Oswego tea
>
> **Parts Used:** All aboveground parts of the flowering plant
>
> **Actions:** Diaphoretic, antifungal, carminative, antiseptic, astringent, and anesthetic

A pungent plant, bee balm is similar in odor and appearance to culinary varieties of oregano. The leaves are lance-shaped, opposite, glabrous or slightly fuzzy, and have a tendency to curve backward toward the ground. The leaf margins are sometimes, but not always, toothed. Like most members of the Mint family, the stems of bee balm are distinctively square. The flowers form in 1- to 3-inch terminal clusters, each containing dozens of tiny, purple- or rose-colored *(M. didyma)* blossoms (they remind me of flames). The size of this plant varies by climate and habitat; plants in the eastern half of the United States are much larger than those of the western states. In the eastern half of its range (where it is commonly known as Oswego tea), bee balm may grow more than 3 feet tall with leaves that can be more than 6 inches long. But in the West, most *Monarda*s are 6 to 18 inches tall, with proportionately smaller leaves and flowers.

Habitat and Range: Bee balm likes meadows and slopes that are predominantly dry and sunny. Like most mints, this plant is adaptable to a wide variety of soils, but unlike most mints, bee balm is drought tolerant and prefers dryland habitats. *M. fistulosa* generally grows below 4,000 feet. Other species of bee balm, such as *M. methaefolia,* grow at higher elevations in shady, moist soils.

The range of this plant is largely undefined, but it seems to be spreading westward from the eastern United States. Livestock enjoy grazing on this plant and have undoubtedly influenced the expansion of bee balm's range.

Applications: Indians of eastern North America have always regarded bee balm as an important medicine. *Monarda* tea is a reliable sore mouth and throat remedy, with a strong antibacterial action that herbalists attribute to the plant's high thymol content (thymol is the active ingredient in many commercial brands of mouthwash). Besides its usefulness as a disinfectant mouthwash or gargle, the plant has mild astringent and anesthetic (slightly mouth-numbing) qualities.

Many contemporary herbalists value bee balm as one of nature's best antifungal medicines. They use it in the form of a tea, tincture, or footbath, or as a vaginal douche to inhibit the growth of *Candida albicans* and various other forms of fungal infection. Bee balm tea is also a diaphoretic (it induces sweating), which herbalists use to help cleanse the body of waste materials via the

Bee Balm   *Monarda fistulosa*

skin. Like most aromatic mints, bee balm tea will help with eliminating gas from the digestive tract. Used as a fresh poultice or in salve form, bee balm is an excellent first-aid dressing for cuts, abrasions, insect bites, and other wounds.

Alternatives and Adjuncts: For use in cold and cough remedies, bee balm combines well with marshmallow root, plantain, yarrow, mullein, coltsfoot, red root (*Ceanothus* species), or arrowleaf balsamroot; the specifics of the infection determine the choice. For first-aid applications, bee balm combines well with plantain, Saint John's wort, poplar buds (balm of Gilead), self-heal, or yarrow. For an antifungal alternative, look to western sweetroot (*Osmorrhiza* species, "sweet cicily" in the eastern half of North America), fireweed, or angelica. Because of its reliable antibacterial activities, bee balm tea is some-times used as a base substance for astringent herbs (such as uva-ursi or pyrola) for treating urinary tract infections.

Propagation and Growth Characteristics: Bee balm is a perennial that reseeds readily and transplants well. Other species of *Monarda* are difficult to start from seed, but *M. fistulosa* is not. This is an excellent plant to introduce into the herb garden. No stratification or other special treatment is required, and it adapts to almost any soil.

Gathering Season and General Guidelines: Gather the leaves, flowers, and stems when the plant is in full bloom (May to September, depending on location). If bee balm is not profusely abundant in your bioregion, selectively pluck a few leaves and flowers from several plants to avoid stripping the stand or gather the stem and leaves after the plant has bloomed and gone to seed; the plant will not be as potent during this stage, but it is still viable for me-dicinal use. Clip the stems about 1 inch above ground level to allow for pe-rennial regrowth and root protection. If you harvest bee balm while the plant is in bloom, always leave several flowers intact for pollination and seed devel-opment. Beware of the possible presence of herbicides in grazed areas, par-ticularly if you notice what your agricultural community may dub "noxious weeds."

Care after Gathering: This herb is generally used as an infusion for sore throats. Dry the leaves and/or the aboveground parts in a dry, dark, well-ventilated area. Though the fresh plant can be used, the herb tastes much more pleasant when dried. The dried herb will keep in quality condition only for six months or so, even if properly stored in airtight containers, so be con-servative when collecting. To make a tea, add a heaping tablespoon of the dried herb to a cup of boiling water.

∿Tincture

Fresh herb: 1:2 ratio in 40 percent alcohol. *Dried herb:* 1:5 ratio in 40 percent alcohol.

Plant-Animal Interdependence: As its name implies, bee balm is highly attractive to bees and other pollinators. This is a reproductive function for bee balm and also for neighboring plants.

Unless more favorable sources of forage have been depleted, bee balm is eaten by wildlife only in limited amounts, therefore this plant is an excellent indicator of forage quality. Assess the quality and quantity of other forage plants in an area before you begin gathering. If wildlife have foraged bee balm heavily, the ecosystem is probably under some form of stress. Check into it, take notes, then move on to a different area.

Tread Lightly: Bee balm grows in abundance where livestock grazes. When gathering from grazed areas, evaluate how seriously the livestock may be damaging the area. Watch an area for several days before wildcrafting—you may be surprised. Where I live, in western Montana, it is common to find a dense stand of plants one day, only to find the same stand completely devoured or trampled by cattle the next.

Black Cohosh
Cimicifuga racemosa

Buttercup Family
Ranuculaceae

Other Names: Black snakeroot, bugbane, squawroot
Parts Used: The root
Actions: Antispasmodic, sedative, nervine, alterative, anti-inflammatory, vasodilator, hypotensive, and emmenagogue

Black cohosh is a large perennial that may grow to 8 feet tall. The white- or cream-colored flowers of this plant are on proportionately long, narrow, tightly clustered terminal racemes that extend more than 18 inches above the rest of the plant. That trait distinguishes *Cimicifuga* from its somewhat toxic neighbor, baneberry *(Actea rubra)*. The racemes develop small, egg-shaped fruits by late summer, each containing numerous flat, brown seeds that resemble little brown bugs. The large (up to 3-foot), coarsely toothed, dark green leaves are pinnately compound in two or three segments. The terminal leaflet of each leaf has three distinct lobes, the middle lobe larger than the other two. The bitter, woody, resinous root of this plant is very dark, earning black cohosh its common name.

Baneberry looks very similar to and is often confused with black cohosh. In comparing the plants side by side (which is how they often grow), the two may be difficult to differentiate if they are not bearing flowers or fruit. Baneberry flowers come in shorter, wider clusters than black cohosh and produce flattened red or white berries, making it easy to distinguish from black cohosh.

Habitat and Range: A native of North America, at least two species of black cohosh inhabit rich, moist woodlands in eastern North America east from Wisconsin and Missouri and south from southern Ontario to Georgia. *C. racemosa* is the most widespread species in that region. In the West, isolated stands of *C. elata* and *C. laciniata* inhabit coniferous woodlands in Oregon and Washington, but the surviving plants are so few that their mention here is so you can honor their survival.

Although black cohosh still grows in abundance in some areas, it is rare in others. Overall, this plant's population is shrinking because of overharvesting and loss of habitat.

Applications: This herb is best known for its use in treating cramping pain related to painful, delayed menses. It contains an estrogenic principle that herbalists believe assists in maintaining hormonal balances in the female reproductive system, and it is used during menopause for hot flashes, tension headaches, and other symptoms. Herbalists also find the antispasmodic and nervine qualities of this plant useful in relieving the pain of sciatica and various forms of neuralgia, and use the plant to ease a spasmodic, nighttime cough. This is likely attributed to the plant's sedative effect on the central nervous system. Scientific studies indicate that the anti-inflammatory principles of black cohosh may be especially useful in treating various forms of rheumatism. Black cohosh has been used for centuries to ease muscle contractions of the uterus during childbirth. It has also been found useful in treating tinnitus.

Alternatives and Adjuncts: Black cohosh combines well with blue cohosh (which in some cases will serve as a substitute) or angelica (especially Dong Qui) for uterine cramping. Wild cherry bark (*Prunus* species) will act as a substitute cough suppressant in many cases. For inflammation and/or rheumatoid conditions, black cohosh is often combined with diuretic herbs, such as dandelion leaf, shepherd's purse, or burdock, or with cholagogues, such as dandelion root or Oregon grape.

Propagation and Growth Characteristics: Black cohosh thrives in rich, moist soil in forest clearings that receive a few hours of slightly filtered sun-

Black Cohosh   *Cimicifuga racemosa*   Black Cohosh flower   *Cimicifuga racemosa*

light each day. These conditions are relatively easy to replicate in the garden, and the live rhizomes of this attractive plant are available through specialty seed companies and some nurseries (see "Resource Guide"). Plant the rhizomes in fall, burying them horizontally under about 2 inches of well-packed soil (press it *firmly* over the root) in a partially shaded location, perhaps under your favorite shade tree or at the north side of your house. Mulch the plantings with plenty of natural leaf material. The rhizomes may take a full year or more to produce an aboveground sprout, so don't be disappointed if the plant does not emerge the following spring. Keep the soil moist, and chances are good that you'll have a beautiful *Cimicifuga* to call your own.

Gathering Season and General Guidelines: I do not condone the wild harvest of this plant unless an immediate and well-defined personal need arises and you have no better alternative. The roots can be harvested in the fall of their fourth or fifth year. If you are harvesting wild roots for emergency purposes, be sure to replant part of the rhizome in the hole you have dug, taking care to use plenty of whatever natural mulch is around after you firmly pack soil over the rhizome (the firming is essential). If you have a home garden, take part of the rhizome for cultivation there. If you are in a patch of wild black cohosh, notice where you put your feet—you might be trampling another (perhaps rare) useful plant.

Care after Gathering: After harvesting the roots, remove excess soil from them with a potato brush or similar tool. If you plan to store the dried roots for an extended period, avoid washing them, which can promote mold and spoilage. Follow the instructions in "Harvesting and Handling Herbs in the Fields" for drying roots. The dried root can be used to make decoctions (see "The Basics of Making Herbal Preparations"), but their resinous nature makes them unsuitable for use in simple infusions (teas). Your best option is to make a tincture of the fresh or dried root.

∾Tincture

Fresh root: 1:2 ratio in 60 to 80 percent alcohol. *Dried root:* 1:5 ratio in 60 to 80 percent alcohol. You will need to find some uncut grain alcohol, such as "Everclear," or a bottle of 151-proof alcohol spirits to reach the alcohol level you need for this herb.

Plant-Animal Interdependence: In areas where black cohosh grows as the predominant forest floor cover, it provides additional shade for a variety of shorter plants, many that are becoming rare. Beneath the canopy of black cohosh's spreading leaves, you might find shade-dependent plants like goldenseal, ginseng, wild yam *(Dioscorea villosa),* or Virginia snakeroot *(Aristolachia serpentaria),* all of which may not survive if the black cohosh is removed. Black cohosh typically grows at the edges of forest clearings or in spots where the overhead canopy allows extra sunlight. These niches are critical to the natural functioning of an ecosystem, as they provide forage and specialized habitats for a diversity of wildlife.

Tread Lightly: Black cohosh is becoming scarce in many areas where it once flourished: the time for commercial-scale cultivation is overdue. As I write this, almost all commercial black cohosh is from the wild. The problems associated with the continued use of wildcrafted black cohosh extend far beyond overharvesting by the herb industry. The delicate habitats where this plant typically grows are quickly disappearing under the pressures of urban development and logging.

The ecological niche where black cohosh flourishes is rich with a diversity of interdependent life. During a recent visit to Appalachia, a group of about twenty students and I counted the plants, insects, and signs of animal presence in a 100-foot square of forest. The plant count alone was more than forty species, and this was in an area where biodiversity is heavily affected by human feet on a regular basis! Be careful where you walk when you're in black cohosh territory, staying on trails whenever possible.

> **WARNING! This herb should be avoided during pregnancy and should be used as a birthing herb only under the direction of a qualified practitioner.**

Blue Cohosh
Caulophyllum thalictroides

Barberry Family
Berberidaceae

Other Names: Blue ginseng, papoose root, squawroot
Parts Used: The root
Actions: Emmenagogue, antispasmodic, uterine-stimulant, anti-inflammatory

Nothing in nature resembles blue cohosh when it emerges in early spring. Unfurling from the earth as a deep-blue stem and a single, folded-up bluish leaf, the plant contrasts vividly with its drab, winter-dormant surroundings. A few days after emerging, yellowish green or greenish purple flower clusters appear, adding to the plant's unique announcement of spring.

The mature 1- to 2-foot plant produces only one large leaf that is thrice-compounded into groups of pinnately arranged leaflets, which are each lobed two or three times, giving them a unique, goosefoot appearance. The entire plant has a bluish hue, as if its smooth, green leaves were sprayed with a transparent blue watercolor. The terminal berries are deep blue when mature.

Although they share similar common names, blue cohosh and black cohosh (*Cimicifuga* species) are not related.

Habitat and Range: Blue cohosh is at home in rich, moist, shady woodlands, from eastern Canada south throughout the eastern third of North America to Arkansas. It extends more sparsely into North Dakota and the central provinces of Canada.

Applications: Blue cohosh is used in much the same applications as black cohosh. The greatest active difference is blue cohosh's weaker antispasmodic qualities and its tendency to elevate blood pressure (black cohosh *lowers* blood pressure). It is traditionally used as a "birthing herb"; its antispasmodic and uterine-stimulant activities reputedly ease false labor pain and improve the efficiency of uterine contractions during labor, making delivery easier. It is regarded as a specific tonic when uterine weakness or lack of smooth-muscle tone has increased the risk of miscarriage. Blue cohosh is sometimes used to help tonify smooth muscle of the urinary tract as well, in the treatment of urinary incontinence. The root has estrogenic qualities, and scientific research suggests that it might act to inhibit ovulation, making it potentially useful as a contraceptive but also complicating its use during pregnancy. The anti-inflammatory and antispasmodic activities of this plant have been validated in clinical and laboratory studies, and its special affinity for the female reproductive system warrants continued research into its uses for uterine inflammation and menstrual cramping.

Alternatives and Adjuncts: Black cohosh is sometimes used as a substitute for blue cohosh, and the two herbs are often combined when stronger antispasmodic actions are required. The fact that these herbs serve similar

Blue Cohosh   *Caulophyllum thalictroides*
Inset: Blue Cohosh flower —Paul Strauss photo

therapeutic purposes benefits both the herbalist and the plants. In many areas where blue cohosh is abundant, overharvesting has depleted neighboring stands of the more popular black cohosh. Both plants are often seen growing alongside one another.

Before using blue cohosh, investigate angelica and motherwort (*Leonuris cardiaca*) as adjuncts or alternatives.

Propagation and Growth Characteristics: Blue cohosh is a slow-growing perennial that can be transplanted from cuttings of its rhizomes. The rhizomes, when planted in rich, firmly packed soil, will quickly take root but may not emerge above ground for two years. Be patient! This plant requires at least 75 percent shade and moist, compost-rich soil with a pH level between 4.5 and 7. The live rhizomes are available through specialty seed companies (see "Resource Guide"). Plants do not reach maturity or peak medicinal potency until they are at least four years old.

Gathering Season and General Guidelines: I do not condone the wild harvest of this plant unless an immediate and well-defined personal need arises and you have no better alternative. The roots can be harvested in fall of their fourth or fifth year. If you are harvesting wild roots for emergency purposes, be sure to replant part of the rhizome in the hole you have dug, taking care to use plenty of whatever natural mulch is around after you firmly pack soil over the rhizome (the firming is essential). If you have a home garden, take part of the rhizome for cultivation there.

Care after Gathering: The roots should be dried in the same manner as for black cohosh. Once again, I would opt to make a tincture, which has an indefinite shelf life, instead of storing the dried root.

✎Tincture
Fresh root, chopped: 1:2 ratio in 40 to 60 percent alcohol (100 proof vodka works fine). *Dried root:* 1:5 ratio in 40 to 60 percent alcohol.

Plant-Animal Interdependence: In areas where blue cohosh still grows in abundance, it is a critical source of cover and concealment for everything from the napping deer to the scurrying field mouse. A thick stand of this plant serves as a secondary protective canopy beneath the trees, helping to prevent soil erosion and damage to ground-hugging plants from driving rain or heavy hail. Blue cohosh is specially adapted to low-sunlight conditions, meaning that it is vital in habitats where many other plants cannot survive. Take blue cohosh out of the picture, and the surrounding habitat will suffer tremendously.

Tread Lightly: Blue cohosh is not as popular in the marketplace as black cohosh, but almost 100 percent of it comes from wild sources. Its slow growth and its requirement for full shade make it particularly vulnerable to overharvest. Reserve this herb for special circumstances, where it is specifically indicated. Then, use it in conjunction with other herbs, dietary adjustments, or other considerations that will allow it to work at an optimum level of efficiency. This plant is a precious healing ally, and until it is cultivated, we must learn to use it consciously, conservatively, and respectfully.

WARNING! Because of the estrogenic and uterine-stimulant activities of this plant, blue cohosh should not be used during pregnancy without qualified supervision. If used in excessive quantities, the powdered root may cause irritation to mucous membranes.

Bugleweed
Lycopus americanus

Mint Family
Labiatae

Other Names: Water horehound
Parts Used: All aboveground parts of the flowering plant
Actions: Hypothyroid, vasoconstrictor, nervine, diuretic, and sedative

The appearance of bugleweed characterizes it as an atypical member of the Mint family. Like other mints it has four-sided stems, flowers at the upper leaf axils, and opposite leaves. But the irregularly lobed leaf shape of *L. americanus* differentiates it from other mints. Even other species of bugleweed (such as the less common *L. unifloris* and *L. asper*) have leaves that are not deeply lobed but simple and coarsely toothed, making them difficult to differentiate from other mints, such as skullcap. The easily identified *L. americanus,* however, is by far the most common and conspicuous *Lycopus* in North America, making it the bugleweed to remember. Bugleweed, unlike many mints, does not have a minty odor. Small whitish to pink flowers are whorled in the leaf axils. The stems may be lightly to moderately hairy, especially toward the base of the plant.

Habitat and Range: An inhabitant of stream banks and marshes, bugleweed typically grows beneath willows and other shrubs. It is widespread throughout most of North America.

Applications: Bugleweed preparations are well known from folkloric accounts as effective cough remedies, which have earned it the common name "water horehound" (derived from another longstanding cough medicine favorite, horehound *[Marrubium vulgare]*). It also has diuretic, hemostatic, and mildly sedative effects when used in tea or tincture form. It is used by modern herbalists for a wide variety of medicinal applications, including treatment of hyperthyroidism, migraine headache, and various nervous disorders. Herbalists believe bugleweed promotes sedative and vasoconstrictor actions that make it useful in many heart and vascular system disorders, working on the cardiovascular system in a way similar to the drug digitalis by strengthening the heartbeat while slowing a rapid pulse. For more in-depth information on the use of this herb, see *Medicinal Plants of the Pacific West,* by herbalist Michael Moore.

Alternatives and Adjuncts: For use in hyperthyroidism, bugleweed is sometimes combined with motherwort *(Leonuris cardiaca).* For use in heart and vascular disorders associated with a rapid, erratic pulse, ginkgo and motherwort will substitute in many cases, and skullcap, hawthorn, garlic, and astragalus should all be investigated as adjuncts. For nervousness associated with abnormal irritability, pain, or rapid heartbeat, bugleweed combines nicely with valerian, skullcap, hops *(Humulus lupulus),* or chamomile. When water

Bugleweed   *Lycopus americanus*

retention is an associated factor in any of the above, try adding dandelion leaf to the formula as a diuretic aid.

Propagation and Gathering Season: Bugleweed requires consistently moist soil and at least 50 percent shade cover. For best results, sow the seeds indoors in February or March, then transplant them to an area where they will have plenty of shade and water throughout the summer. Plants are ready to harvest when the flowers are just beginning to bloom, usually late June to early August.

Care after Gathering: The freshly cut herb is the best choice for making your own medicine, but if this isn't possible, cut the stems at about 3 inches above ground level during a period of dry weather. Remember: this is a plant that typically likes a wet environment, so the herb is especially prone to mold while it is drying. Hang small bunches of four to six stems in an airy location out of direct sunlight. When properly stored, the dried herb will retain its medicinal potency for at least six months.

∾Tincture
Fresh herb: 1:2 ratio in 40 to 60 percent alcohol. *Dried herb:* 1:5 ratio in 40 percent alcohol.

Plant-Animal Interdependence: Although seldom foraged and not particularly attractive to pollinating insects, bugleweed contributes to its environment by helping to break up the continuity of the soil with its rhizomatous root system. Bugleweed also provides low ground cover for amphibians, insects, reptiles, and small mammals, especially where it predominates in thick profusion beneath the shade of willow, red osier dogwood, or other riparian-area shrubs and trees.

Tread Lightly: Bugleweed typically grows in riparian thickets that are especially sensitive to damage by humans. These biodiverse niches are critical safe havens for countless organisms who live together to form a unique and complex microecosystem. Human presence alone can be disruptive enough to cause serious imbalance.

Although this plant is under no current threat of overharvesting, it is difficult to justify an incursion into a sensitive habitat to collect even a small personal supply of this plant. If you collect wild bugleweed, do so from the margins of the thicket. Stay alert, and if you see or hear a bird or animal alarmed by your presence, move on to a less sensitive stand. Better yet, grow bugleweed in your garden or buy it from a cultivated source.

WARNING! Bugleweed should not be used during pregnancy.

Burdock
Arctium species

Other Names: Clotbur, lappa
Parts Used: Mainly the root, sometimes the seeds
Actions: Alterative, diuretic, and bitter

A biennial, burdock first emerges as a rosette of large, irregular, basal leaves. The second year produces a thick, heavily leafed stalk that can grow to 10 feet tall. The leaves grow alternately from the erect stalk, which eventually develops alternate, leaf-bearing branches. The large 3- to 12-inch-wide basal leaves of the mature plant are heart-shaped. The upper leaves are smaller and vary in shape to nearly oval. The lower surfaces of all the leaves are paler than the upper surfaces. The entire plant is covered with fine hairs that make the plant feel sticky-abrasive—like extrafine-grit sandpaper.

The terminate flowers are purple and form on bristly, thistlelike burrs. Each burr is equipped with a multitude of backward-hooked spines, which attach to anything that brushes against them. After the plant matures and dries, the burrs harden, making the hooks even more effective. Each burr contains several small, elongated seeds.

The root is large (often huge!), tapered, and slightly aromatic. You may need a shovel, digging bar, and axe to extricate the root, which can be 3 feet long and weigh several pounds.

Burdock is often confused with its relative, cocklebur *(Xanthium strumarium)*. The difference between these two plants is in their burrs. On burdock, the burrs grow from the stem tips, but on cocklebur the burrs grow along the stems. Also, the burrs of burdock contain several dark, small seeds,

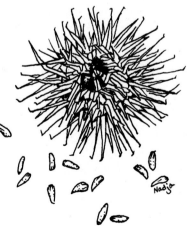

Gobo Burdock *(Arctium lappa)* seedpod

and the burrs of cocklebur contain two flat seeds that look like shelled sunflower seeds.

Habitat and Range: Burdock grows in disturbed areas, such as roadsides, landfills, irrigation ditches, and alluvial accumulations. A native of Europe and Asia, burdock has been introduced into most temperate regions of the world. It grows sporadically at low elevations, usually below 5,000 feet, across North America.

Applications: Herbalists worldwide know burdock as a "blood purifier." It does not directly "clean" the blood, but gently stimulates the liver to function more efficiently, thus helping the organ with its burdensome job of filtering waste materials from the blood, making new red blood cells, and producing bile, enzymes, and other constituents critical to the metabolic processes of nutrient absorption and waste elimination. Burdock is highly nutritious, containing large amounts of iron, a vast array of minerals, and vitamins A, C, D, and E. All parts of this plant are edible. The root can be sliced and used in a salad or lightly cooked . . . it is very tasty.

In both Western and Eastern modalities of herbalism, burdock is used as a specific, long-term treatment for eczema, oily-flaky scalp, and other disorders the herbalist may associate with liver or digestive deficiency. It is also believed to possess antioxidant qualities, protecting the body (at a cellular level) against the potentially harmful effects of carcinogens and various toxins. For use in these applications, liberal doses of a tincture preparation or root tea are usually recommended.

Alternatives and Adjuncts: Dandelion, yellow dock, red clover, milk thistle, or licorice root combine well with burdock, and in many cases are equal or better substitutes.

Propagation and Growth Characteristics: A biennial, burdock is inconspicuous its first year, then reaches for the sky during its second and final growing season. After blooming, the plant dies, leaving behind thousands of seeds to be distributed by anything that walks by. It is a highly successful survivalist. Burdock is still marketed as a vegetable in Eurasia. Although it probably was introduced into North America for the same purpose, it is now generally regarded as a troublesome weed that invades pastures, fields, and waste areas. Burdock grows especially well in drainage areas where soil accumulations are rich and deep, and where livestock or human traffic affords plenty of opportunities for seed transport.

Once this plant has become established, it is likely to proliferate. It prefers sunny locations with rich soil and plenty of moisture but will grow with some success under almost any conditions.

Gathering Season and General Guidelines: Although the root is a usable medicine anytime, fall harvest of first-year growth is recommended because digging a mature burdock taproot can be like excavating a short telephone

Burdock   *Arctium minus*

pole. Another consideration is the burr factor. They'll cover you! If you don't mind the burr attack, be sure to remove every last little beastie from your clothing and hair before moving on to another locality: burdock can be an aggressive intruder to the unaccustomed ecosystem.

If you are collecting the roots to eat, gather only young, first-year plants, which are tender, mild, and easy to work with; old plants are tough and bitter.

This is an easy plant to grow in the garden, but pull the plants the first year, harvesting the root in the fall, to avoid any risk of escape.

Care after Gathering: Use the root fresh or dried; I like to use it fresh, which eliminates shelf life problems and the loss of active constituents. Cut the roots into manageable pieces, lengthwise then crosswise, before drying, tincturing, or infusing. Dry the roots with good air circulation.

⌒Tincture

Fresh root: 1:2 ratio in 50 percent alcohol. *Dried root:* 1:5 ratio in 50 percent alcohol. *Seeds:* 1:5 ratio in 50 percent alcohol.

Plant-Animal Interdependence: Burdock is an alien that invariably alters the ecosystems where it flourishes, but as it conquers and claims a new neighborhood, it offers some positive attributes. The massive roots effectively reduce erosion in high-impact areas, and its abundant foliage provides food and shelter for birds and small animals in areas where habitat may already have been damaged by human-related activities (such as excavation or free-ranging livestock). When the juicy leaves of burdock die back, they contribute substantial quantities of rich compost to the soil.

Though large animals seldom eat it, burdock is a rich food source when indigenous food plants have been depleted. Evidence of animal forage on this plant indicates an ecosystem under stress.

Tread Lightly: Because burdock is aggressive, the wildcrafter must respect its potential for restructuring an ecosystem, beneficially or otherwise. The longer burdock has been in a particular ecosystem, the more defined its role becomes. As Earth's caretakers, it is important for us to recognize, respect, and understand symbiotic relationships—even if those relationships involve an introduced plant. We must rethink the collective attitude that labels burdock a "troublesome weed" and consider the possible consequences of attacking this plant with herbicides where it has settled in as a member of the biocommunity.

Take extra care when handling this plant to assure that you do not inadvertently introduce it into pristine areas. One burr can make the difference between a chemical-free environment and one that has been poisoned with chemical herbicides.

 **Catnip**
Nepeta cataria

Other Names: Catmint
Parts Used: All aboveground parts of the flowering plant
Actions: Mild sedative, carminative, digestive antispasmodic, and
   diaphoretic

Catnip is unmistakably a mint, with square stems and opposite leaves. The
entire plant is covered with soft, velvety hairs that give the plant a blue gray
appearance. The leaves are ovate, ½ to 2 inches wide, with distinctively round-
toothed margins. The flowers bloom both in terminate clusters *and* from the
leaf axils of the upper plant, on proportionately long peduncles, a characteristic that sets catnip apart from most other mints. The ¼- to ½-inch-long
tubular flowers are white with purplish pink spots. Wild catnip is much
stronger than the commercially grown varieties used in feline entertainment
products. The fresh plant has a strong, skunky-mint odor that demands rec-
ognition and is very different from the stale, dried product from pet stores.
Plants range from 6 inches to 5 feet tall when mature, depending on growing
conditions.

Habitat: Catnip was introduced from Europe with the earliest Anglo coloni-
zation. It was regarded as a "cure-all" and has been distributed to nearly every
part of the globe that was explored by European colonists. It prefers sunny,
moist, disturbed areas and frequently grows along roadsides, irrigation ditches,
pastures, and rangelands across North America.

Applications: Although you will not experience the euphoric qualities from
catnip that your cat might, catnip tea is well known by herbalists as a relaxing, after-work tea. Like most other mints, catnip will help relieve the body of
excess gas, and it imparts a mild sedative-antispasmodic action to a nervous
stomach. Catnip is a very gentle sedative; it does not induce any sensation of
intoxication but helps calm the herbalist's stomach after he realizes he's too
old to combine beans, lard, hot salsa (with too much garlic), heavy metal
rock, and a review of the monthly phone bill in a one-hour period.

Alternatives and Adjuncts: The dried leaves and flowers make a pleasant-
tasting catnip tea to which any combination of herb tinctures may be added
as needed. For indigestion, gas, or a generally nervous stomach, chamomile
or pineapple weed is a weak alternative to catnip. If all you need is a remedy
for intestinal gas, field mint, peppermint, or virtually any other aromatic mint
will fill the bill.

Propagation and Growth Characteristics: Catnip is a perennial that likes
rich, slightly acid, well-drained soils with full sun and ample moisture. It can
be propagated from seed but is easier to establish from transplants or root
cuttings. Catnip is a hardy plant, adaptable to northern climates, and is

Catnip *Nepeta cataria*

commercially grown with great success as far north as Alaska. Although an alien, catnip is not highly competitive with other plants. It is an excellent choice for the home herb garden.

Gathering Season and General Guidelines: Gather catnip just as it is beginning to bloom, generally in midsummer. Selectively collect leaves or clip stems. If you gather the stems and upper plant, cut the stem 6 inches or more above ground to protect the roots and ensure perennial growth and leave plenty of flowers for pollination and natural seed distribution.

Care after Gathering: Dry the herb as soon as possible according to the instructions in "Harvesting and Handling Herbs in the Field." Toss the herb frequently to expedite the drying process and help prevent mold. Dried catnip is best used as a tea and is very pleasant tasting, especially with some added honey. The fresh plant, although useful, tends to be strong and bitter tasting.

Plant-Animal Interdependence: Catnip grows in thick, bushy stands that provide habitat for birds, amphibians, and reptiles. It is an effective pollinator attractor and hosts a multitude of insects and spiders. I have seen birds and the occasional deer eat catnip, but it is not high on the herbivore menu. Mountain lions, lynx, and bobcats occasionally cut loose and partake in the plant's feline narcotic qualities, but not to an extent that the wildcrafter needs to worry about stumbling upon a big, tripped-out pussycat.

Tread Lightly: Catnip sometimes grows in soils that compress easily, so wear soft footwear and gather the herb during dry periods. Catnip typically grows among a proliferation of other plants, where you'll have to maneuver carefully to avoid harming neighboring plants and organisms. Try to gather from the periphery of large patches; removing too much vegetation may compromise insect and small animal habitat. A visual difference after harvest means you took too much.

Chickweed
Stellaria media

Pink Family
Carophyllaceae

Other Names: Common chickweed, starweed
Parts Used: The whole, fresh aboveground plant
Actions: Emmolient, demulcent, refrigerant, and diuretic

Chickweed is a sprawling weed that typically grows in dense mats along streams, in irrigation ditches, and in pastures. Several species are widely distributed across North America.

The leaves of chickweed are opposite, ovate to broadly lanceolate, and up to 1½ inches long. The flowers are simple and white, with five, two-part petals that give each flower a ten-petal appearance. The flowers grow from leaf axils on the numerous branch tips.

The stems have distinctive, minute hairs down one side. They are characteristically weak and angular, each with a single line of diminutive hairs along one angle and with root nodes at intervals that enable the plant to creep, like strawberry runners.

The entire plant is sweetly juicy and is a delicious addition to salads or sandwiches.

Habitat and Range: If you haven't already discovered this plant, you need only look a little closer. Chickweed was introduced into North America from Eurasia and has made itself at home from coast to coast. Its wide distribution in the United States is largely from agricultural and landscaping activities. It is common to moist pastures, lawns, vacant lots, and roadside ditches.

Applications: A fresh chickweed poultice cools and sooths minor burns, insect bites, and other skin irritations. Fresh chickweed can be infused into olive oil to make a lotion for dry, chapped skin, or for use in salves and ointments. Internally it does the same thing—it cools and soothes irritations of the mouth and esophagus. If eaten in large quantities, it can act as a laxative.

Alternatives and Adjuncts: Chickweed adds a soothing element to first-aid oils and salves, where it combines particularly well with Saint John's wort, comfrey, hound's tongue, bee balm, or balm of Gilead (poplar). Alternatives to chickweed include cleavers and aloe vera (aloe, by the way, is far more versatile than chickweed).

Propagation and Growth Characteristics: Common chickweed is an annual that likes ample moisture and at least half a day in the shade. It is easy to grow and reproduces readily. The plant crawls around, putting down roots along the way, while producing seed throughout its growing season (which can be all year in mild climates). Some of the seed companies that offer "gourmet salad green" plants are selling chickweed seed for the home garden. My crop came to me in the form of horse manure! Although reproductively

Chickweed   *Stellaria media*

aggressive, chickweed is not highly competitive with other plants and is easy to control in the garden provided you harvest it frequently.

Although it grows in any soil, chickweed prefers rich loam with a neutral to slightly acid pH. The soil must be consistently moist but not boggy. Because chickweed continually drops viable seed during its perpetual bloom season, transplanting this herb is as easy as dropping a handful of the fresh, flowering greens into a consistently moist flowerbed.

Gathering Season and General Guidelines: Chickweed can be harvested anytime. If you are gathering it to eat raw, beware of plants that are wet and growing close to water—they might be harboring *Giardia* or other nauseating microorganisms. Always wash and dry the greens before eating them.

Despite its delicious flavor and medicinal attributes, chickweed may grow among troublesome weeds. Be aware of the possible presence of herbicides, particularly in cultivated areas. Avoid gathering this plant from roadsides or city lots, where the plants tend to collect toxic residues.

Care after Gathering: Chickweed should be used fresh. As it dries, it loses valuable protein constituents quickly, so it is best to make tinctures and oils in the field. The fresh herb is also useful and pleasant as a tea.

∾Tincture
Fresh herb: 1:2 ratio in 40 percent alcohol.

To make an oil infusion, let the chickweed wilt for a few hours after harvest; by removing some of the water from the leaves and stems, you will minimize the risk of mold and greatly extend the shelf life of the finished product. After the herb is wilted (but not dried), stuff it into a glass jar and completely cover it with olive oil. Put a lid on the jar and place it in a warm area in the house for about one month. Then, using a sieve or ricer, press and strain out the oil through a paper coffee filter, muslin, or some cheesecloth. Add a few liquid drops of vitamin E to each ounce of the oil as a preservative, and store the oil in a clean jar in the refrigerator. The oil should keep for several months.

Plant-Animal Interdependence: Animals find chickweed as delicious as we do. In areas where the plant dies back each fall, the green, juicy leaves and stems compost readily and are high in plant nutrients. Chickweed generally indicates rich soil.

Tread Lightly: Be careful to avoid introducing this plant into areas where it may not be welcomed by the "noxious weed" patrol. Stands that appear to be a food source for wildlife should be left alone—animals need the plants more than we do, and there is plenty of this abundant herb for all of us. Find a healthy stand that is easy to access where you will cause minimal damage. Soft, wet stream beds can and should be avoided.

Cleavers
Galium species

Madder Family
Rubiaceae

> **Other Names:** Bedstraw, catchweed, goosegrass, cleavers wort, clivers
> **Parts Used:** All aboveground parts of the fresh plant
> **Actions:** Diuretic, anti-inflammatory, astringent, vulnerary, and lymphatic

Galium is a large genus, with eight species native to the Rocky Mountain bioregion alone. All share a similar appearance.

Galium species typically have leaves that are narrow, ½ to 2 inches long, growing in whorls of four to eight that radiate from the stem in intervals, like the spokes of a bicycle wheel. The stems are four-sided. The flowers are tiny, white, and star-shaped. The entire plant is delicately fragrant, especially when bruised, with a sweet, rose odor. Until the early twentieth century, the plant was a popular choice for mattress stuffing (one of its common names is "bedstraw"). Of the several *Galium* species, *G. aparine*, known as "cleavers," is the one herbalists use most often. Although all *Galium* species are medicinally useful, herbalists prefer *G. aparine* because of its juicy, water-soluble nature.

Cleavers *Galium aparine*   Northern Bedstraw *Galium boreale*

The name "cleavers" probably originated from *G. aparine*'s distinctive ability to cling to things—particularly clothing—the same way hair sticks to something with a static charge.

Cleavers is a weak-stemmed plant that grows in tangled, climbing masses in moist, usually riparian areas. The plant has tiny, downward-pointing hooks along the angles of its stems, which enable it to cling and climb on virtually anything. The tiny hooks are easier felt than seen, making the stems feel sticky, especially when you run your fingers upward along the stem.

The seed capsules of cleavers look like pairs of little green balls. Herbalist Michael Moore aptly describes them as "covered with little bristly hairs like green testicles."

Cleavers is a delicate annual that grows in moist, easily compressed soils that are usually host to a thick profusion of other plants. Other species, such as *G. boreale* (northern bedstraw), grow in drier, more impact-resistant habitat. Northern bedstraw closely resembles cleavers except that it grows on stout, self-supporting stems and does not have the clinging characteristic of cleavers. And unlike cleavers, northern bedstraw is a perennial, which means it is much easier to monitor for regrowth and the effects of harvesting.

Habitat and Range: Cleavers grows in moist, partially shaded areas—along streams, in wet draws and irrigation ditches, and on moist, north-facing hillsides. It likes to climb and is typically tangled with neighboring plants. Northern bedstraw and dozens of other perennial species are common to drier, sunny areas. All *Galium*s like rich soil and are widely distributed throughout North America.

Applications: Herbalists regard cleavers as one of the best herbs for assisting with the drainage of lymph-engorged tissues. They credit it with mild vasodilating qualities that work especially well to open tiny capillaries that are swollen, abscessed, or engorged with excess fluids. Cleavers also has mild diuretic qualities, which may help rid the body of excess fluids.

Herbalists frequently use cleavers in the healing of stomach ulcers, ovarian cysts, tonsillitis, or in circumstances where the lymph circulation seems to be chronically or acutely impaired. Because this herb is safe in large doses over extended periods, it is commonly used as a preventative "lymphatic tonic."

Externally, cleavers is useful for treating insect bites, stings, and other irritations and ulceration of the skin. The cooled, diluted tea is sometimes used as an anti-inflammatory eyewash.

Alternatives and Adjuncts: As a specific remedy for tonsillitis, red root (*Ceanothus* species) may prove more useful than cleavers in acute situations. For the skin, plantain, self-heal, nettle, chickweed, and Saint John's wort should be considered as an adjunct or alternative to cleavers. Nettle and raspberry leaf are excellent eyewash alternatives. Oregon grape root or usnea lichen combine well with any of those herbs for producing an antibacterial.

Propagation and Growth Characteristics: Cleavers, an annual, has delicate roots, does not transplant well, and is nearly impossible to start from seed. Northern bedstraw, however, and many other perennial *Galium* species, are effective reproducers that easily take hold from transplants or seed. Stratification or other special treatments are unnecessary, and you can easily introduce the plant into the herb garden. Several garden cultivars of *Galium* are available through most nurseries, sold under the common name "sweet woodruff."

All *Galiums* require consistently moist soils with neutral to slightly acid pH levels; most will do best if they receive at least a few hours of shade each day.

Gathering Season and General Guidelines: Harvest cleavers and other *Galiums* during the early stages of the blooming period, typically between mid-May and July. Cleavers dies back immediately after blooming; if you need some, don't procrastinate harvesting it. Clip off northern bedstraw and other perennial species above ground level to allow for next year's growth.

The annual cleavers uproots easily, as its root system is weaker than the stem. This can be a problem for wildcrafters; unlike perennial species, the reproductive success of annuals such as *G. aparine* depends entirely on seed distribution. If our harvest kills too many plants, next year's stand may be compromised. To ensure sufficient seed distribution, carefully snip off the upper third of the plants you select and gather only from the outer edges of healthy stands.

Care after Gathering: Most herbalists agree that the best way to prepare and use cleavers is to press the juice from the fresh plants and consume it in raw form (several juicers are suitable for the job). The juice spoils quickly— try pouring it into an ice cube tray and freezing it. You can pop the cubes out for use as needed. Because cleavers must be used immediately after harvest or made into a fresh plant tincture (it loses much of its valuable protein and enzyme constituents when it dries), it is a good candidate for tincturing at the site of harvest. Have a jar and alcohol menstruum at hand when you gather the plant.

∿Tincture

Fresh herb: 1:2 ratio in 50 percent alcohol; cut the plants with scissors and put them in a jar with the alcohol.

Plant-Animal Interdependence: *Galium* tastes as sweet as it smells. It is relished by everything from field voles to the always hungry moose. Birds and small animals build nests in the tangle of vines that cleavers creates.

Tread Lightly: If you enter cleavers' delicate habitat, be aware of vulnerable soil structures and the possible presence of bird nests and other animal dwellings. Move carefully and try not to damage other plants when working within a tangled clump of cleavers. Wear soft footwear and monitor the effects of your presence closely for two or more years after gathering. Annuals must distribute seed each year to survive, so harvest conservatively.

Clematis
Clematis species

Buttercup Family
Ranuculaceae

Other Names: Virgin's bower, old man's beard, pepper vine, pipestems, traveler's joy
Parts Used: Leaves, stems, and flowers
Actions: Diaphoretic, diuretic, and vasoconstrictor

Clematis is a climbing perennial, with strong, vinelike woody stems that typically entwine around the branches of neighboring trees and shrubs. Clematis may cover an entire stand of shrubs but go unnoticed until it blooms. Several clematis species, both native and introduced, grow in North America. *C. columbiana,* a common native species, is best recognized by its showy lavender flowers, which instead of petals presents four tapered sepals that look like petals—a little paper lantern. After blooming, the flowers develop into silvery white, seed-bearing plumes. The leaves are opposite on long leaf stems (petioles) and are divided into three pointed leaflets, each with one side indented and a rounded base. The climbing stems may reach 8 feet long. The native *C. ligusticifolia* produces small, inconspicuous, cream-colored flowers, though the overall plant is large (stems up to 20 feet). *C. virginiana,* an eastern North America species, also has white flowers and closely resembles *C. ligusticifolia.*

Habitat and Range: *C. columbiana,* one of the most widespread western woodland species, grows on wooded hillsides, particularly in Northwest coniferous forests at subalpine elevations, from British Columbia to Oregon and east into Montana and Wyoming. *C. ligusticifolia,* by far the most common lowland species, is widespread across the western half of North America in streamside thickets and ravines from below sea level in the deserts to about 4,000 feet in the New Mexico mountains and north to British Columbia.

Applications: Although clematis earned the common name "pepper vine" because early travelers used it to spice up salads, the entire genus (and most of the Buttercup family) contains strong chemical constituents that can irritate skin and mucous membranes. Ingestion can cause internal bleeding. Many references list this species as poisonous, even though Native Americans for years used it as a remedy for sore throats and colds. Its unique vasoconstrictor-dilating action may account for its use by advanced practitioners in treating some types of migraines.

Physicians once used a liniment of this plant for treating nervous disorders and skin eruptions, but mainstream medicine no longer uses it. Today, careful use of this plant is believed to trigger a unique set of immediate actions on the surface capillaries of the brain, and experienced herbalists are using it at the onset of migraine headaches.

Anything that offers a breath of hope against the debilitating pain of migraine has to be our friend, even if it can be harmful. Until we learn more about the medicinal use of this plant, it will remain one of many potential

Clematis flower   *Clematis columbiana*

Clematis flower   *Clematis ligusticifolia*

Clematis leaf   *Clematis columbiana*

sources for healing that live in the forest. By appreciating clematis and other plants for their unknown potential, we learn to recognize the deeper values of the forest—and with appreciation comes willingness to protect and heal the wildlands that sustain us.

To read more about the therapeutic usefulness of this plant, see Michael Moore's classic book, *Medicinal Plants of the Mountain West*.

Alternatives and Adjuncts: Feverfew *(Tanacetum parthenium)* or periwinkle *(Vinca major* or *V. minor)* can be used for long-term treatment and prevention of migraine headaches. You should be able to find both at your local plant nursery. Other alternatives include bugleweed *(Lycopus americanus* or *L. virginicum),* culinary ginger, or cayenne.

Propagation and Growth Characteristics: Clematis, a perennial, produces viable seeds that generally need cold, moist stratification to germinate. Most species prefer shade and must have something to climb on. *C. columbiana* sometimes is available through nurseries, particularly ones specializing in native plants. Clematis is a beautiful addition to a shady trellis. It is slow to start from seed, though, and does not transplant well from the wild.

Gathering Season and General Guidelines: Clematis is most potent after the flowers have finished blooming, just before the leaves begin to die back for winter—and when the overall appearance of the plant resembles poison ivy! To assure positive identification, find a healthy and accessible stand of flowering plants in early spring. Make notes of landmarks and the plants' leaf and stem characteristics, then plan on returning in late summer or fall to harvest. To harvest, clip the end 12 to 24 inches of the vinelike stems with a sharp pair of pruning clippers. Be conservative about how much you take, and gather from a variety of different plants to disperse the effects of your harvest.

Care After Gathering: The leaves and stems lose their medicinal potency shortly after they are harvested, so it's best to make your tincture in the field. With your pruning clippers, cut the leaves and stems as finely as possible, place them in a glass jar, and cover them immediately with alcohol menstruum. Be sure the lid is on tight before you put the jar into your backpack.

∾**Tincture**
Fresh herb: 1:5 ratio in 50 percent alcohol.

Plant-Animal Interdependence: The strong, often mat-forming stems of clematis provide habitat and nesting material for birds and other small animals. The long-lasting flowers are effective and important pollinator attractors.

Tread Lightly: When studying this plant on steep slopes, wear adequate footwear and use a walking stick to avoid slipping and unnecessarily damaging the site. If you venture into thick brush, be aware of the possible presence of small animals and their dwellings.

WARNING! This plant contains compounds that may cause internal bleeding if ingested in large quantities.

Cow Parsnip

Heracleum lanatum

Parsley Family

Umbelliferae

Other Names: Cow cabbage, American masterwort, wild parsnip, woolly parsnip, wild celery, Indian celery

Parts Used: The dried roots or the ripe (but not dried) seeds

Actions: Digestive antispasmodic, carminative, and anesthetic

Like most members of the Parsley (Umbelliferae) family, cow parsnip has tiny, five-petaled, white flowers arranged in umbrella-shaped clusters (umbels). Mature plants produce several of these umbels, but the one growing from the center stalk is usually the largest. After blooming, the umbels produce large (up to ¼-inch) seeds that are flat on one side and slightly rounded on the other, each with distinct ridges that alternate with four black lines. The leaves are very large, often growing to 12 inches or more, and are divided into three parts, like a maple leaf. The leaf margins are serrated. The stems are stout and hollow, somewhat hairy, and grooved along their entire length. The roots are large, tapered, light brown, and usually have a pungent, carrotlike odor.

Cow parsnip is one of the largest members of the Umbelliferae family, often reaching more than 7 feet tall. Many people shy away from cow parsnip because of its relationship to poisonous water hemlock *(Cicuta douglasii)*, but this plant is easy to distinguish from its deadly cousin. Unlike water hemlock, the leaves of cow parsnip are very large and not at all carrotlike. The robust appearance of cow parsnip makes it stand out from all other members of the Parsley family.

Habitat and Range: Cow parsnip likes wet, boggy soil and grows along streams and springs and in wet meadows. It seldom strays more than a few feet from a reliable water source and requires at least partial shade and deep, rich soil. Cow parsnip is widespread across North America, from sea level to about 9,000 feet.

Applications: Although not frequently used medicinally, cow parsnip has a long history as an effective remedy for digestive and nervous disorders. Herbalists sometimes use the roots to treat various gastric disorders, especially abdominal cramping associated with nervousness, indigestion, and/or diarrhea. The root of this plant is acrid and strong tasting (especially when fresh), so most herbalists opt for a more pleasant-tasting alternative.

In my experience, the seeds are the most useful part of this plant. When chewed or used in an infusion, they are anesthetic to the mouth and gums, similar to the effects of clove oil. The numbing effect is not strong enough to address deep-tissue pain, such as an abscessed tooth, but it is useful for temporary relief of minor irritations, such as that pretzel you chewed with your gums by mistake. The seed tea can also be used for stomach upset, diarrhea, and gastric cramping.

Cow Parsnip   *Heracleum lanatum*

Alternatives and Adjuncts: For gastric upset and cramping, chamomile, pineapple weed, or virtually any mint is gentler than cow parsnip. For generalized abdominal cramping, most herbalists would turn to black cohosh or angelica before using this herb.

Propagation and Growth Characteristics: A native perennial that reproduces by seed, cow parsnip is effectively distributed by seasonal runoff. It can be propagated from seed if stratified and kept very moist throughout germination. The plants grow rapidly, producing a large taproot requiring rich, deep soil that must remain consistently moist.

Gathering Season and General Guidelines: Collect the seeds during late summer, after they are ripe and the dark stripes have become clearly visible. Dig the roots after seeds have fallen from the umbels. Do not gather any part of this plant if it cannot be positively identified—deadly poisonous water hemlock often grows nearby cow parsnip.

Although this plant is edible (trust me—it tastes terrible!), it has a reputation for causing photosensitive contact dermatitis in some people, so wear long sleeves and gloves when collecting the seeds or roots.

Care after Gathering: The roots of this plant are too strong and acrid to be used fresh—they may *cause* severe digestive upset rather than *cure* it. Using a heavy knife, split or cut the roots once or twice into lengthwise quarters and let them dry with plenty of ventilation. The seeds can be used fresh or dried.

ᥭTincture
Dried root: chopped or ground, or *Fresh seeds:* 1:2 ratio in 50 percent alcohol.

Plant-Animal Interdependence: Deer, elk, moose, and bears forage on cow parsnip throughout its growth period. As the name implies, cows sometimes eat it, too.

Cow parsnip umbels are important pollinator attractors, drawing insects from the forest canopy that later move on and pollinate other understory plants. A stand of large, leafy cow parsnip provides an important source of shelter and food for insects and small animals and sometimes a cool, midday bed for large animals. Shade and moisture-loving companion plants, such as pyrola or cleavers, grow at cow parsnip's feet.

The juicy leaves and stems of this plant decompose quickly each fall, adding substantial quantities of organic matter to the soil. The large root plays a natural role in soil aeration and preventing erosion.

Tread Lightly: The deleterious effects of gathering this herb vary with each micro-ecosystem. If you collect the plant from mucky bogs and junglelike growth, your damage is likely to be high. Look for a healthy stand that is easy to access, such as those bordering a well-traveled foot trail. Incursions into vulnerable areas are unnecessary given the extensive range and abundance of cow parsnip. Inspect prospective stands for evidence of active foraging—

nibbled leaves, animal tracks, or animal droppings—and if such signs are present, move on to another stand. Gather this plant only during dry periods (but not drought), and wear soft-soled footwear to minimize soil compaction.

> **WARNING!** This plant may cause photosensitive contact dermatitis in a small percentage of individuals. If you anticipate physical contact with this plant, proceed with caution.

Cranesbill Geranium
Geranium maculatum and *G. viscosissimum*

Geranium Family
Geraniaceae

Other Names: Wild geranium, alumroot, sticky geranium, storksbill, cranesbill, geranium mack, doveroot, shameface, old-maid's-nightcap

Parts Used: The entire plant, especially the roots

Actions: Astringent and hemostatic

Also known as "sticky wild geranium" because of the texture of its foliage, cranesbill geranium is typically seen as a predominant summer show of bright pink in dry grasslands across temperate North America. The palmate leaves each have five distinct lobes and form proportionately long leaf stems (petioles) that extend directly from the base of the plant. The showy, five-petaled, light pink to purplish red flowers have distinctive hairs on their inner surfaces, and the sepals and flower stalks are covered with yellow-tipped hairs. Cranesbill geranium may grow as high as 32 inches. The plant is named for its long, beaklike fruit capsule, a universal characteristic of the Geranium family.

Habitat and Range: Cranesbill grows in open grasslands, where it shares habitat with arrowleaf balsamroot, lupine, sagebrush, and other dryland plants. To a lesser degree it grows in moist areas as well, up to timberline. *G. viscosissimum* is the predominant species in the western third of the United States and Canada. *G. maculatum* predominates in the east, from Manitoba south to Georgia and east into Missouri and Kansas.

Applications: Cranesbill geranium is useful in a wide variety of situations where an internal or external astringent is indicated. Its broad spectrum of usefulness is because its medicinal potency is adjustable, according to which part of the plant you use. The leaves and flowers are mildly astringent, making them suitable for treating minor skin irritations or in an anti-inflammatory eye rinse. The root of the plant is much more potent and is useful for treating acute cases of diarrhea, cystitis, or gastrointestinal bleeding. Dried and ground into a powder, the roots can be applied to razor cuts and minor lacerations as a styptic agent. Made into a tea, the leaves, flowers, or roots can be used as a gargle for a sore, inflamed throat, or for swollen gums. Best of all, the entire plant is safe to use.

Alternatives and Adjuncts: For stronger astringent actions, turn to alumroot (*Heuchera* species) or uva-ursi. Effective substitutes for cranesbill geranium include goldenrod, pyrola, and plantain. Combined with marshmallow root (*Althea officinalis*), cransebill geranium is an impact-resilient alternative to slippery elm (*Ulmus fulva*), a tree at risk of becoming endangered because of overharvesting and the effects of Dutch elm disease.

Propagation and Growth Characteristics: Cranesbill geranium can be started from transplants or seed. To collect the seed, wait until the plant has

Cranesbill Geranium   *Geranium viscosissimum*

Cranesbill Geranium   *Geranium viscosissimum*

finished blooming and the bill-like fruits are just beginning to dry. Pick a few of the fruits, place them on a dish, and allow them to dry completely. The seeds they contain should be viable and can be sowed in fall in a sunny location. Once established, this plant is cold hardy and drought tolerant and will flourish in just about any soil. Cranesbill geranium blooms throughout summer, adding vibrant color to areas that are otherwise dry and drab. Potted plants are available through nurseries specializing in native landscape plants.

Gathering Season and General Guidelines: The fresh leaves, stems, and flowers of this plant can be used anytime. So can the roots, but they are much more potent if dug in the fall, after the foliage has begun to die back. The leaves and flowers can be gathered with a minimum of damage to the ecosystem, but harvest of the root will usually result in the death of the plant. With this in mind, try using the flowers and leaves before turning to the root. If you dig the root, try snapping the top of the root off and replanting the root crown in the hole.

Care after Gathering: Chop up fresh aerial parts (the stems, leaves, and flowers) and infuse them with boiling water. Chop and decoct the fresh roots, or dry for later use (see "Harvesting and Handling Herbs in the Field").

Plant-Animal Interdependence: The bright pink flowers of cranesbill geranium attract pollinating insects, and hungry deer and rodents sometimes browse the foliage. This plant commonly grows on dry hillsides and embankments that are nearly void of other plants, where it is effective in controlling erosion.

Tread Lightly: Many areas where this plant grows suffer from heavy cattle grazing or recreational vehicle activities. The wildcrafter must recognize the ill effects he or she may be contributing to an already compromised ecosystem.

Dandelion
Taraxacum officinalis

Sunflower Family
Compositae

Other Names: Chicoria, lion's tooth, piss-in-bed
Parts Used: The spring leaves, the fall roots, and the flowers anytime
Actions: Diuretic, cholagogue, nutritive, bitter, and laxative

Do you really need me to describe this plant? Surprisingly, it is misidentified more than most self-respecting herbalists care to admit. The main point to remember when identifying dandelion is that the plant has no leaf stems. The leaves grow in a rosette, directly from the rootstock. Dandelion has no true stems; the only parts that resemble a stem are the peduncles—the leafless stalks that bear individual, terminate flowers. If the plant has leaves presented from a central *stalk,* it is not a dandelion. Also, if the plant has any branching characteristic, then, once again, it is not dandelion. Of the plants confused with dandelion, spotted cat's ear *(Hypochaeris radicata),* sow thistle *(Sonchus* species), and the wild lettuces *(Lactuca* species) are among the most common. When young, all these plants look very much alike. None of these impostors harm you if ingested, but neither will they yield the same medicinal results as dandelion. Wait until the plants are mature enough to identify before you harvest.

Habitat and Range: No surprises here: the dandelion grows just about everywhere, from below sea level to above timberline. It likes full sun and grows with less flower development in shade. It tolerates any soil but prefers rich, well-drained loam with neutral acidity. You'll be hard-pressed to find country that does not have a dandelion growing in it.

Applications: For people who wish to harvest, process, and use their own herbal medicine, I can think of no better introduction than dandelion. This plant offers an incredibly broad spectrum of therapeutic and nutritional usefulness. When dandelion was introduced from Europe, it was known as "piss-in-bed herb" because the diuretic actions of the leaf tea are so strong. Dandelion is unique as a diuretic because it is a rich source of potassium and other trace minerals and vitamins, which replace what would otherwise be lost through the urination it stimulates. For this reason, dandelion is widely used by herbalists in the treatment of any variety of disease that involves fluid retention, such as pulmonary edema or gout.

Dandelion is considered a safe but powerful liver stimulant and is frequently used to treat various liver and digestive disorders, as well as chronic skin conditions that may result from an overworked or dysfunctional liver or gall bladder. As a "bitter," fresh dandelion leaf or leaf tincture is useful if consumed immediately before eating a meal to "kick start" the digestive process before food enters the stomach, which maximizes digestion and nutrient absorption. The flowers are a rich source of lecithin, and herbalists credit them with mild analgesic qualities.

Dandelion   *Taraxacum officinalis*

The beauty of dandelion as an herbal medicine is that it contains diverse active constituents in safe quantities that the body can freely and completely assimilate. It is the perfect example of a "plant food medicine"—rich in protein, inulin, iron, and potassium, and vitamins A, C, and B-complex (to name a few). From a holistic perspective, dandelion is indicated for use under almost any circumstances. It is in the largest bottle in my medicine chest, and it seems to show up in most of what I formulate for myself or my companion animals.

Propagation and Growth Characteristics: Dandelion is a perennial that grows as an annual in severe climates. Although this plant is regarded as a weed across North America, dandelion seed is available through some commercial seed suppliers. Surprisingly, it is a slow starter in the garden, but once it establishes itself . . . well, you know.

Dandelion is a popular salad green in Europe and is gaining popularity in the United States. Hybrid food strains of this plant are available in the produce sections of some natural food stores.

Gathering Season and General Guidelines: The best leaves are plucked while green and juicy, as the plant is beginning to bud, but in a pinch they can be gathered anytime. Dig the roots in late fall, when they contain the greatest concentration and diversity of useful constituents. You might plan your root-digging expeditions for when the soil is damp and soft, as the long taproots can be difficult to extricate from hard, dry soil. I like to dig the large second- or third-year roots after the first autumn rain. A round-nosed ditch shovel is effective for digging this herb, as the roots are stubborn.

Because this valuable plant annoys so many people, always beware the possible presence of herbicides. Do not gather dandelion from roadsides, agricultural waste areas, cultivated fields, public parks, or golf courses. In fact, don't gather it at all, unless you are positive that the area has not been sprayed with an herbicide. Herbicide residues are sometimes present in an environment three or more years after application. Although banned from use in the mid-1970s, residues from dichlorodiphenyltrichloroethane (whew! that's DDT) were still showing up in soil samples more than twenty years later. Be careful out there!

Care after Gathering: The leaves and roots can be used fresh or dried. If you are using the leaves for tea, dry them first to improve palatability. *Do not wash them* or mold may result.

For tincturing, the fresh herb is best, but dried leaves or roots will yield a good end product.

⌒Tincture

Fresh root or leaves: 1:2 ratio in 45 percent alcohol. *Dried root or leaves:* 1:5 ratio in 45 percent alcohol.

Plant-Animal Interdependence: Dandelion generates a wealth of mineral-rich composted matter, is an abundant source of early-season food, and is a bright yellow incentive for pollinators. The persistent taproots effectively break up the hardest of hardpan soils and often play critical roles in erosion control. Dandelion is good at surviving insect infestation and drought.

Tread Lightly: Dandelion is a "free love" kind of plant, which distributes its seeds on the wind, by runoff, and through animal transport. When you gather dandelion, be careful to avoid transporting seed into pristine environments. Many local, state, and federal agencies, as well as individuals, wage chemical warfare against this herb. You can help reduce the spread of chemical pollution by aggressively harvesting dandelions from areas that have not yet been sprayed. Local weed control agencies sometimes lead those efforts. If the plants or plant parts you have collected include flowers or seeds, carry them in a covered bucket or other container that won't lead to unintentional distribution of seeds.

Echinacea
Echinacea species

Sunflower Family
Compositae

Other Names: Purple coneflower, black Samson, rudbeckia, Missouri snakeroot

Parts Used: The mature roots (at least three years old) or the flowering tops (a weaker, but useful medicine)

Actions: Immunostimulant and antimicrobial

Echinacea is a naturally long-lived perennial that ranges from 6 to 40 inches tall, depending on species and growing conditions. A dark, distinctively cone-shaped flower center characterizes the genus. The 2- to 5-inch-diameter flowers range from pale purple pink to deep purple (depending on species), with the petals drooping downward as the plant matures, adding to the conelike appearance. The shape of the flower identifies echinacea as an obvious member of the Sunflower family.

The leaves are lanceolate to nearly oval. *Echinacea purpurea* leaves are toothed, but many other species are not.

Habitat and Range: Echinacea's rapid decline in the wild is in no small part because of habitat loss. A native to prairies, open meadows, and sunny woodland clearings, echinacea once grew in abundance from the continental divide eastward. *E. angustifolia* and other prairie species once flourished in the northern plains of Wyoming, Nebraska, and the Dakotas, and where species such as *E. pallida* once thrived in the central corn belt and the South, populations have been drastically affected by the advent of large-scale commercial farming and overharvesting by the natural products industry.

Applications: Echinacea is an effective, time-proven immune system tonic that was widely used by Native Americans. Recent scientific validation and effective media-marketing campaigns have turned echinacea into an herbal sensation. Today, it may be the most well-known herb in the world but also one of the most misused.

These days, echinacea is showing up in everything. I recently saw a bottle of "energy boosting" soda that listed it as an ingredient; even some shampoos now contain echinacea! Ludicrous! Echinacea works within very narrow parameters but is touted as something that does not exist—an all-purpose herb. It will assist a healthy immune system to work more efficiently, but it does *not* act as an herbal substitute for a depressed immune system. It does *not* increase a person's energy level, at least not beyond its "feel better" capacity to help the body ward off microbial infection. Echinacea is a valuable, effective herbal medicine, but like all other herbs, it has definite limits to its usefulness.

What echinacea does best is this: it provides a needed auto-immune boost at the early onset of bacterial or viral infection. Echinacea works best to help

Echinacea   *Echinacea purpurea*

ward off a cold or flu when administered *before* the infection establishes itself in the body—when you feel that first little tickle in the back of your throat. Although echinacea can help prevent a viral infection at the time of exposure (when everyone in your office is coughing), it definitely is not a nutritional supplement for everyday use, "just in case." Echinacea used that way can make the body adjust to the prolonged presence of the herb, negating its immune-stimulating actions.

Alternatives and Adjuncts: In the western third of North America, arrowleaf balsamroot serves as a good bioregional alternative to echinacea. If you are in the desert Southwest, read about using yerba mansa *(Anemopsis californica)* in Michael Moore's book *Medicinal Plants of the Desert and Canyon West,* as an alternative.

Virtually any of the alterative diuretic, cholagogue, or expectorant herbs combine well with echinacea to support the body against various forms of microbial infection. To help fight infections of the mouth, digestive tract, or urinary tract, echinacea is an excellent adjunct to Oregon grape root or usnea lichen.

Propagation and Growth Characteristics: *E. purpurea,* the primary echinacea of commerce, is easy to grow and requires little care once the plants are established. The seeds require cold, damp stratification and light to break their dormancy. This means that the seeds must be sown on top of the soil or covered with just a trace of soil to germinate. Species such as *E. pallida* and *E. angustifolia* are finicky but not impossible to grow in the garden. Before you end up pulling your hair out trying to grow these species, though, remind yourself that despite what herbalists once thought, the gardener-friendly *E. purpurea* provides a medicine as potent as any others.

Plants are hardy to at least zone 4 and are strongly drought tolerant when mature. A well-established stand of echinacea will reseed itself, and mature plants can bear roots of up to 3 pounds each. Roots must be at least three years old to be of medicinal value.

Some nurseries sell young echinacea plants as perennial landscape flowers. Check with your local suppliers for species and cultural information suited to your area.

Gathering Season and General Guidelines: Dig mature echinacea roots from the garden when the plant has gone dormant for the winter, after its third year of growth. The leaves, stems, and flowers can be selectively harvested when they are in full bloom, in midsummer. If you opt for that, remember that the plants will need a preponderance of leaves and flowers left on them to provide life-support for the roots and to produce viable seeds.

Care after Gathering: The flowers, leaves, and roots can be dried and used for tea, or tinctured. To dry and store the roots, leave them whole until you are ready to use them, to help maximize shelf life. Properly stored roots will keep for two or three years.

∼Tincture

Fresh, chopped roots: 1:2 ratio in 70 percent alcohol. *Dried roots:* 1:5 ratio in 70 percent alcohol. *Flowers and leaves:* 1:5 ratio in 70 percent alcohol.

Plant-Animal Interdependence: Echinacea is an important pollinator attractor in its native environment. Animals and insects often eat the plants (and gophers love the roots), perhaps in part for the same reasons we do.

Tread Lightly: Habitat loss and the greedy behavior of too many wildcrafters is rapidly wiping out wild stands of echinacea. In the south, east, and central United States, commercial wildcrafters sometimes eradicate wild echinacea as quickly as it is discovered.

Please do not wildcraft this herb. Instead, use echinacea products made from cultivated sources. The survival of this beautiful flower depends on our support of organic farmers—and a large measure of luck for the plants that remain in the wild. If you find a wild stand, keep it a secret and protect it. It may be the last field of wild echinacea you ever see.

Elderberry
Sambucus species

Honeysuckle Family
Caprifoliaceae

Other Names: Elder, red elder, blue elder, black elder, sambucus
Parts Used: The flowers and berries
Actions: Diaphoretic, diuretic, expectorant, an effective laxative, and
possibly antiviral qualities

Elderberry is a shrub or many-branched small tree that can grow to 25 feet tall, but is generally smaller at high elevations or in severe climates.

There are several species of elderberry. Most people differentiate them by berry color. The blue and black species are commonly gathered for jams, pies, and elderberry wine. The red species have a reputation for being toxic, especially the seeds, leaves, bark, and roots.

Elderberry leaves are pinnately compound, growing in two to four opposite pairs of lance-shaped 1- to 5-inch leaflets, with a single leaflet growing from the tip of each compound leaf. The leaflets generally have saw-toothed margins with sometimes hairy undersurfaces.

The flowers are in dense, upright clusters about 4 inches across. Individual blossoms are white to cream colored, have five petals, and are each about ¼ inch wide. The flowers are fragrantly sweet and developed into clusters of berries, often so abundantly that they weight down the branches.

Elderberry is sometimes confused with mountain ash (*Sorbus* species), a shrub or small tree that has similar leaves and clusters of bright red berries. The leaves of mountain ash tend to be more oval than those of elderberry and have a serrated edge with the points all directed toward the tip of the leaflet. Mountain ash leaves are more uniform in shape and smaller than those of elderberry and are often alternate instead of opposite. Elderberry leaves are very large and curl backward, like peach leaves. Once the wildcrafter has identified elderberry and mountain ash in the field, the distinguishing characteristics usually become obvious.

Habitat and Range: Elderberry grows in the moist soils of mountain hillsides and along roadsides and stream banks. It frequently inhabits old clearcuts and burns where it gets plenty of sunshine and potassium-rich soil. Several species are common in the United States, at elevations ranging from sea level to nearly 10,000 feet.

Applications: Elderberry's recent claim to fame stems from some promising clinical research suggesting that extracts of the berry may inhibit the reproduction of various strains of influenza. Several scientifically refined preparations of elderberry fruit are now available on the market for this purpose. The self-reliant herbalist might obtain similar results from a good tincture of the ripe berries, but as of this writing, documented cases of success are still few and far between.

Elderberry fruits *Sambucus cerulea*

Elderberry *Sambucus cerulea*

Elderberry has a long history as a cold remedy herb. Until the recent scientific hoopla, this herb was best known as an effective diaphoretic and expectorant, which herbalists find useful for moving mucus out of the upper respiratory tract, and for inducing sweats during a dry fever.

Alternatives and Adjuncts: Yarrow combines well with elderberry as a diaphoretic-antimicrobial-expectorant adjunct. Arrowleaf balsamroot adds immune modulating and expectorant actions. Coltsfoot further stimulates the expectorant action of elderberry.

Propagation and Growth Characteristics: Elderberry is a deciduous shrub that likes ample moisture and at least partial sun exposure. You can start it from seed, but it's a slow grower. Wild shrubs do not transplant well at any stage. Many fruit tree nurseries sell elderberry trees that are relatively fast and easy to establish in a sunny spot in your yard. In its natural habitat, elderberry seeds are distributed by animals that eat the berries.

Gathering Season and General Guidelines: Gather the flower clusters when the buds are just beginning to open, usually between mid-May and July. Berries are ready when they are fully ripe in mid- to late summer, but collect them before they begin to dry up. Gather only flower clusters and berries within easy reach, to help minimize damage to the tree and surrounding habitat.

Care after Gathering: Use the flowers either fresh or dried in tea, or for food recipes. Cook and strain fresh elderberries for jams. The flowers and berries can be dried for future medicinal use, but shelf life is limited to six months or less. Dry them on a nonmetallic screen, butcher papers, or in paper bags with the tops left open. Avoid exposing them to sunlight during and after the drying process.

∾Tincture

Fresh, ripe, mashed berries: 1:2 ratio in 50 percent alcohol.

Plant-Animal Interdependence: Elderberry is an important food plant for birds and many mammals. It provides a popular nesting habitat for our winged companions and is an effective pollinator attractor during its profuse bloom period. Sometimes you can detect this plant by the sound of bees humming around the flowers! Where elderberry grows along roadsides or in burns, it is important in habitat regeneration and erosion control.

Tread Lightly: This abundant herb should be gathered conservatively from several different locations and stands. Be aware of the possible presence of small animal and bird dwellings in and around elderberry branches. Do not gather from stands where active animal foraging is evident, especially where bears have been feeding. Bears strip elderberry shrubs of every juicy berry they can reach, prompting the earth-conscious herbalist to wonder what a bear might do if it found me raiding its elderberry patch.

WARNING! Species of elderberry that produce red fruits may be toxic. The seeds and bark of the entire *Sambucus* genus contain hydrocyanic acid, a substance that may be toxic if ingested in large quantities.

False Solomon's Seal

Lily Family

Smilacina racemosa and *S. stellata*

Liliaceae

Other Names: Solomon's plume, wild lily of the valley
Parts Used: The root preferably, and also the upper plant parts
Actions: Demulcent, anti-inflammatory, mild antitussive, astringent expectorant, and sedative

A very pretty inhabitant of the forest floor, false Solomon's seal has rhizomes that creep through moist forest compost and emerge as dark green, 8- to 32-inch leafy stalks. Two species grow in North America, both similar in appearance. False Solomon's seal is a perennial with long, creeping root stalks from which solitary, branchless stems emerge in early spring. The leaves are alternate and lance-shaped, with the base of each leaf partially clasping the stem (if the leaves clasp *completely* around the stem, it's not false Solomon's seal). Only two species of *Smilacina* inhabit the West. Both have small, white, starlike flowers that appear at the very top of the stem. *S. racemosa,* the larger of the two (up to 32 inches tall), has broader, 2- to 5-inch-long, prominently veined leaves and has numerous flowers arranged in a densely clustered raceme. *S. stellata* has narrower, generally shorter leaves (2 to 3 inches), and the plant is smaller overall (up to 24 inches tall), with only a few flowers presented in a zigzag inflorescence. Both produce reddish berries after blooming.

Both species are sometimes confused with two other similar-looking lilies—twisted-stalk *(Streptopus amplexifolius)* and true Solomon's seal *(Polygonatum biflorum),* which is native to the eastern third of North America. These two look-alikes can be distinguished from false Solomon's seal by their flowers, which oddly hang from the leaf axils of the plants. Twisted-stalk has stems with a more defined zigzag pattern and leaves that wrap completely around the stem of the plant. The leaves of false Solomon's seal grow more directly off the stem. False Solomon's seal flowers are always terminal—they bloom only from the stem tips.

Habitat and Range: False Solomon's seal likes soil rich in organic matter. It is a common plant in shady coniferous forests, from sea level to nearly 10,000 feet, across much of North America.

Applications: Herbalists do not use this plant often, which is just as well, considering the small size of the useful root. Harvesting a usable quantity of this herb usually requires the death of several plants. It's best to think of false Solomon's seal as "a useful option."

Herbalists use the fresh root tea or syrup for easing dry, raspy coughs that produce a raw sore throat. The expectorant actions of this herb help to soften and release mucus buildup in the bronchi, while its mucilage constituents (a slippery-oily substance) help lubricate and protect dry, irritated membranes.

False Solomon's seal contains sitosterol, a compound with proven anti-inflammatory qualities, and allantoin, a compound that speeds regranulation

False Solomon's Seal   *Smilacina racemosa*

(the healing process) of wounds, burns, and especially ulceration in the mouth. False Solomon's seal is a good herb to remember for field applications; chew the root and leaves to relieve sore gums and such, or mash them with some water to make a soothing poultice for insect bites, sunburn, scrapes, or minor bouts with poison ivy.

False Solomon's seal was used by American Indians as a mood elevator. It was ingested or used as a smudge to "ground the spirit" and to bring inner peace.

Alternatives and Adjuncts: For stronger expectorant actions, mullein, colts-foot (*Petasites* or *Tussalago* species), grindelia *(Grindelia squarrosa)*, and arrowleaf balsamroot are better options than false Solomon's seal. For relief of minor injuries, bites, and irritations of the skin, plantain, chickweed, or fireweed are all options to consider. In syrup, false Solomon's seal combines well with arrowleaf balsamroot and horehound.

Propagation and Growth Characteristics: False Solomon's seal is a perennial that grows from laterally creeping roots. The plant is difficult to start from seed, with a double dormancy requiring two years (and two stratification periods) for seeds to germinate. The plants require mainly shaded areas with soil that retains moisture well. Rhizome cuttings are sometimes successful when started in a clean, damp medium (such as peat moss), but the odds are against success.

Gathering Season and General Guidelines: This herb can be dug anytime, but you may want to wait until flowers are present to alleviate any identification doubts. Take the upper plant and only 3 inches or less of the rhizome, which is small and easily damaged. To avoid breaking the stem off above ground level, grasp the rhizome just below the soil and cut it with sharp scissors, leaving plenty of the root for next year's regrowth.

Care after Gathering: Although the root is best used fresh in cough syrups, it can be dried (use standard drying procedures outlined in "Harvesting and Handling Herbs in the Field") and stored for up to six months for use in tea.

To make a syrup, mix the chopped roots with an equal quantity of honey and place it in a glass double boiler. Slowly simmer the mixture until most of the moisture from the plant material has evaporated; be patient—this may take hours. When the mixture is reduced almost by half, strain it through a sieve. If it is too thick to strain, add enough brandy, rum, or your favorite spirit to thin it to a thick syrup. Stored in the refrigerator, the syrup should keep for a year or more.

Plant-Animal Interdependence: The rhizomatous nature of this plant makes it an effective soil aerator. The rhizomes also help hold together mats of coniferous debris, which act as moisture-retaining mulch for the forest floor biocommunity. The extensive root system of this plant creates a subterranean highway for insects, small rodents, and microorganisms.

Tread Lightly: Although this herb is abundant and widespread, the effects of harvesting it are disproportionately high. Harvest false Solomon's seal only when the forest floor is dry and resilient. Avoid soft, mossy, consistently moist areas and places where the plant grows among a dense proliferation of other plants.

Fireweed

Evening Primrose Family

Epilobium angustifolium

Onagraceae

Other Names: Willow herb, willow weed, blooming Sally
Parts Used: The entire flowering plant
Actions: Laxative, demulcent, anti-inflammatory, antispasmodic, and astringent

Fireweed is well known for its spires of brilliant pink flowers that lend bold contrast to burns and other disturbed areas. The bright flowers are four-petaled and terminate, with spire-shaped clusters atop erect stems.

The leaves are narrowly lanceolate, 4 to 6 inches long, and bear a resemblance to many willow species (although this plant is unrelated to the *Salix* genus). The leaves are pale underneath, darker on top, and grow alternately in what looks like an upward spiral along the main stem. The veins of the leaves have the distinct characteristic of joining together in loops at the leaf margins, a peculiarity that distinguishes fireweed from most of its look-alike neighbors and simplifies identification at any stage of its growth.

The fruits are 2 to 3 inches long, very slender, and stand rigidly off the stem. The skinny pods eventually split to release airborne, seed-bearing puffs of fine, white hairs.

Two species of fireweed are common in Northwest mountains, *E. angustifolium* and *E. latifolia*. Both species are similar in appearance, may grow together, and are identical in use.

Habitat and Range: Fireweed inhabits the edges of burns, clearcuts, and other disturbed areas throughout the coastal and inland mountains of Canada and the western United States.

Applications: Fireweed tea is a pleasant, mild laxative. Although it is not used often, some herbalists turn to it for relieving food-induced constipation. A fresh herb poultice will bring topical relief to contact dermatitis, sunburn, and insect bites. A decoction of the aboveground parts is useful as an antispasmodic in the treatment of whooping cough and asthma. The young leaves and shoots are delicious in salads or as a potherb.

Alternatives and Adjuncts: For topical applications, chickweed, plantain, or false Solomon's seal may be useful alternatives to fireweed. As an anti-inflammatory, licorice root (*Glycyrrhiza* species), yucca (*Yucca* species), willow, or poplar are all stronger options. For a stronger laxative effect, look to elderberry, yellow dock, or Oregon grape.

Propagation and Growth Characteristics: Fireweed is a perennial that propagates from airborne seeds and spreading rhizomes. It readily establishes itself along the predominantly downwind edges of burns and other forest clearings. Fireweed is not picky but will do best in deep, well-drained soil

Fireweed   *Epilobium angustifolium*

Fireweed leaf   *Epilobium angustifolium*

high in potassium (wood ash is a good source). It prefers full sun for at least four hours a day.

The seeds germinate easily after a period of cold, moist stratification, with a few of the less common species requiring light to break their dormancy. Root cuttings can be transplanted with success if kept constantly moist until they become established.

Gathering Season and General Guidelines: For medicinal use, gather fireweed while it is blooming (June through September). Cut the stalk above ground level, taking care not to disturb the roots. If you plan to use the root, take only a few inches of the rhizome and leave the rest for perennial regrowth.

For food use, gather young shoots and leaves before the plants bloom in early spring. Although older plants are edible, the leaves become bitter and tough with age—go for the more tender leaf tips and side shoots. Consumption of too much fireweed will probably lead to dramatic overuse of toilet tissue.

Care after Gathering: Although this herb is best used fresh, the roots can be dried for use in salves or in powder form. The aboveground parts should be chopped and used fresh in tea.

Plant-Animal Interdependence: Fireweed is one of nature's earth regenerators. In burns or other disturbed areas, fireweed is one of the first plants to offer a new lease on life for resident wildlife. Fireweed means reliable food and habitat for birds and other small animals, and a napping place for larger animals, which also browse the foliage. Fireweed rhizomes help loosen compacted soils, and the juicy leaves and stems add rich compost to the soil after they die back each fall.

Tread Lightly: For effective habitat regeneration, please do not gather fireweed from areas that have been recently damaged by fire or other serious disturbance. Leave young stands to get on with their job, and find another stand in a more well-established habitat. When gathering in clearcuts, burns, or other disturbed clearings, always be aware of the potential for erosion. Plan your harvest carefully and never gather plants in or near a watercourse.

Goldenrod
Solidago species

Sunflower Family
Compositae

Other Names: Blue Mountain tea, boheatea, wound weed
Parts Used: All aboveground parts of the flowering plant
Actions: Astringent, anticatarrhal, diuretic, carminative, diaphoretic, and tonic

Goldenrod, a common wayside weed, is easy to identify by its terminal, spire-shaped or triangular clusters of tiny, bright goldenrod yellow flowers. The alternate leaves of the most common species are narrowly lanceolate and may or may not have serrated edges. The plants are erect and range anywhere from 2 inches high (*S. multiradiata,* mountain goldenrod) to 70 inches tall (*S. canadensis, S. occidentalis,* and *S. gigantea*). Most species, especially the larger ones, share similar appearances and may differ only in leaf texture or the presence of stem hairs.

Habitat and Range: The *Solidago* genus can be divided into two categories: the ones that grow in moist soils and the ones that prefer dry habitats. Most of the smaller mountain species typically grow in dry soils, often at the edges of forest roads or in open meadows. The larger species typically grow in riparian habitats, irrigated fields, and drainage ditches from below sea level to about 4,000 feet. Several species are widespread across North America.

Applications: Used topically, dried goldenrod in powered form is good for stopping the bleeding of minor cuts and scratches. The most common form for using goldenrod is tea brewed from the dried leaves and flowers. Goldenrod tea may relieve hay fever symptoms if used before onset, but people who are allergic to goldenrod should avoid it. Herbalist David Hoffmann writes, "Goldenrod is perhaps the first plant to think of for upper respiratory catarrh, whether acute or chronic. It may be used in combination with other herbs in treatment of influenza." Herbalists believe that goldenrod helps relieve the inflammation of upper respiratory membranes, reducing mucus production. For this reason, herbalists often use it to move bronchial congestion out of the body.

Goldenrod is also said to be useful in the treatment of urinary tract infections, kidney stones, and arthritis. Some herbalists believe it effectively treats urinary incontinence by tonifying tissues of the bladder and lower urinary tract.

Alternatives and Adjuncts: For treatment of bronchial congestion, goldenrod combines well with coltsfoot, mullein, grindelia, marshmallow, fireweed, or yarrow. For urinary tract applications, consider mullein root, uva-ursi, marshmallow, plantain, fireweed, and pyrola as possible adjuncts. As a styptic agent, yarrow and alumroot are excellent alternative herbs.

Goldenrod   *Solidago canadensis*

Propagation and Growth Characteristics: Most goldenrods are perennials that reproduce from rhizomes or seed. Although many goldenrods grow in the poorest of soils, most prefer deep, rich loam and are easy to grow in your garden. Goldenrod seed is available through retailers, or look for seed companies and nurseries with wildflower seeds (see "Resource Guide").

This plant is considered a "troublesome weed" by some people, but it is not highly competitive with other plants. It typically grows in small patches, often dotting a field or adding sporadic bursts of color to a stream bank—but, as usual, there are exceptions.

Gathering Season and General Guidelines: Gather this plant while it is in full bloom (August to September). Clip the top 6 to 12 inches with a sharp pair of shears, leaving ample plants for reseeding and pollinator activity.

Goldenrod often hangs out with a crowd of weedy friends that have bad reputations. If you are planning to harvest from areas where "noxious weeds" are present, be sure they haven't been sprayed with herbicides.

Care after Gathering: Dry goldenrod for use as a styptic powder or for tea. This plant dries well if hung in loose bunches (no more than 1 inch wide at the bases) covered with a single sheet of newspaper or a paper bag. The paper will protect the plants from dust and sunlight without restricting air circulation. Coarsely crush the dried herb for tea or grind it fine for use as a styptic powder.

∿Tincture

Fresh herb: 1:2 ratio in at least 40 percent alcohol. *Dried herb:* 1:5 ratio in at least 40 percent alcohol.

Plant-Animal Interdependence: Goldenrod's main claim to fame in nature is its ability to attract pollinators from miles around. The sweet flowers also attract ants and other insects, and sometimes deer.

Tread Lightly: While consumer demands are seriously compromising populations of such "trendy" herbs as osha and goldenseal, goldenrod stands undisturbed. It is a very useful herb—in many instances where echinacea or goldenseal are used out of context, goldenrod would likely be a better choice. Goldenrod is abundant, it is prolific, and it's in our backyard—let's use it.

Goldenseal
Hydrastis canadensis

Buttercup Family
Ranunculaceae

Other Names: Yellow root, eye balm, yellow paint, ground raspberry, wild turmeric

Parts Used: The root primarily and the leaves to a lesser extent

Actions: Antimicrobial, anticatarrhal, tonic, and astringent

Goldenseal is a perennial that may grow to 12 inches high. The main stem of the plant typically forks to produce two nearly circular 2- to 6-inch-wide leaves, one usually larger than the other. The leaves are deeply lobed five to seven times, with toothed margins. The plant may take three or more years to bloom, and then it produces a single, whitish green flower that instead of rays has sepals arranged in a concentric cluster. In midsummer, the flower develops into a single, raspberry-like red fruit that contains ten to thirty small seeds. The stems are hairy. The rhizomatous root is thick and woody, with several small rootlets branching away from the main root stalk. All parts of the root have deep goldenrod yellow inner tissues.

Habitat and Range: The range and population of this North American native are rapidly diminishing. The original range included most of eastern North America, from Minnesota and Vermont all the way south into Georgia. Today, most remaining stands of wild goldenseal are isolated in the central and northern reaches of the Appalachians and, less often, the Ozarks.

Applications: Like its neighbor, American ginseng *(Panax quinquefolius),* wild goldenseal is vanishing from the landscape because of sensationalism and misinformed use. The greatest misnomer about this plant is that it acts as an herbal antibiotic in the body, coursing its way through the body in the bloodstream to attack any pathogenic microbes in its path. Goldenseal does not act like an antibiotic. The antimicrobial alkaloids of this plant (berberine and hydrastine) inhibit bacteria, fungi, and parasites that they come in direct contact with only in the digestive and urinary tracts. These infection-fighting compounds are not absorbed into the bloodstream, they are merely contact disinfectants. Another recent untruth about goldenseal is that it will mask the results of a drug screen urinalysis. Nonsense.

The responsible herbalist uses goldenseal to treat mucosal inflammations of the mouth, upper respiratory tract, eyes, and, to lesser extent, the digestive and urinary tracts. It can be used as an effective surface antimicrobial, too, and all these actions can be accomplished with the use of various other herbs that are far less compromised in the wild. If you use goldenseal, please use it only from cultivated sources. Wild goldenseal is one of the most endangered medicinal plants in North America, and it is being exploited by bad apples of the herb industry.

Goldenseal   *Hydrastis canadensis*

Alternatives and Adjuncts: In many cases where an antimicrobial or anticatarrhal is indicated, several plants containing the yellow alkaloid berberine can be used instead of goldenseal. Think about trying Oregon grape, twin-leaf *(Jeffersonia diphylla)*, or yerba mansa *(Anemopsis californica)*. Before using any of these herbs, though, ask yourself a fundamental question of holistic responsibility: By using a wildcrafted substitute for an overharvested herb, am I benefiting the future of wild medicinal plants, or am I simply deferring the consequences to another species? Humans are very efficient at disrupting and depleting natural resources on Earth; when we abandon one natural resource for another, we are perpetuating the continuum. We should use wild substitutes only when cultivated goldenseal is not available. *Coptis sinensis,* a species of goldthread widely cultivated in China, may be another excellent substitute for goldenseal.

Propagation and Growth Characteristics: If you have a piece of ground that will support goldenseal, please grow some! The best place to plant goldenseal is under the shade of a dense, hardwood canopy on a north-facing hillside. It requires deep, compost-rich, well-drained soil with a pH between 5.5 to 6.5. By using shade cloth and appropriate soil amendments, you can grow it in the garden, too.

Propagate goldenseal from stratified seed or from rhizomes spaced 4 inches apart in rows 12 inches apart. Plant in the fall. Goldenseal requires at least four to five years (preferably seven) to reach harvest maturity. For more detailed information, contact one of the resources in the "Resource Guide".

Gathering Season and General Guidelines: The gathering season for wild goldenseal is NEVER. If you find some plants in the wild, keep them secret and help by protecting them. If you have some in cultivation, harvest a few of the leaves and seeds (for planting) every fall until the roots mature. The leaves can be used fresh to make a tincture that is weaker than the root but nevertheless useful. When you harvest roots, be sure to reserve some of the rhizome for transplanting.

Care after Gathering: The roots can be dried whole or used fresh. Properly dried and stored roots will remain in good condition for several years. I prefer to eliminate the shelf-life guesswork by tincturing them.

~**Tincture**

Dried roots, ground: 1:5 ratio in 40 percent (80 proof) alcohol. *Fresh roots:* 1:3 ratio in 40 percent (80 proof) alcohol.

Plant-Animal Interdependence: Nothing I know of eats this plant, but in the few areas where it still flourishes, goldenseal grows as a dense, 1-foot-high ground canopy that provides refuge for nesting birds, reptiles, amphibians, and small mammals. These rare, tightly spaced plant communities are elders of the forest and have been contributing compost and shelter to their ecosystem for several decades. If they are suddenly removed, we lose far more than a precious medicinal plant. These special sites represent what an eastern hardwood forest should be. Such places should be used as classrooms to relearn the natural ways of our planet.

Tread Lightly: Fifty to sixty million goldenseal roots are harvested from the wild each year, but the term *ethically wildcrafted goldenseal* is, by most accounts, an oxymoron. Please don't buy wildcrafted goldenseal—support the farmers who cultivate this precious herb. Cultivated goldenseal is expensive, but the time has come to pay the price to save a precious friend and medicinal ally. Become involved in efforts to save wild goldenseal from land development and logging; rescue plants from areas slated for highway construction; join United Plant Savers (see "Resource Guide") and help educate other people about the need to protect plants like goldenseal. The survival of this plant probably rests with people who are willing to get involved.

WARNING! Goldenseal should not be used during pregnancy.

Goldthread
Coptis species

Buttercup Family
Ranuculaceae

Other Names: Coptis, western goldthread, yellow thread, canker root
Parts Used: The root primarily, and also the upper plant parts
Actions: Antimicrobial, antiviral, cholagogue, laxative, astringent, anti-inflammatory, uterotonic, antiparasitic, and hemostatic

Goldthread is a low-growing, evergreen perennial, typically the predominant ground cover in old-growth forests.

The leaves are divided by three wiry leaf stems (petioles), each bearing a shiny leaflet that looks like a cross between a strawberry leaf and a small oak leaf. The leaflets are ovate, three-lobed, sharply toothed, and glossy on the upper surface. Each leaflet is usually less than 1½ inches wide. The entire plant is rarely more than 6 inches high.

The flowers of goldthread grow at the ends of simple, leafless stems that extend away from the base of the plant. The one to three whitish flowers have five to eight narrow sepals ¼ to ½ inch long, and five to seven narrower petals. When the petals fall off, the remaining green sepals and subsequent fruits look like odd, green, star-shaped flowers. The hollow seed-bearing capsules spread apart at maturity.

The roots of goldthread are thin and typically threadlike, with bright, goldenrod yellow inner tissue. The roots are long, weak rhizomes with the bitter taste characteristic of all herbs containing high amounts of the alkaloid berberine.

Habitat and Range: *C. occidentalis* is common to shady, coniferous forests throughout northern Idaho and the mountains of central California, eastern Washington, Montana, and British Columbia. It likes undisturbed stands of cedar, yew, and grand fir, where it has ample shade and a thick accumulation of forest debris. It grows predominantly at elevations above 2,500 feet but seldom higher than 7,000 feet.

Other species of goldthread, such as *C. groenlandica,* grow in eastern North America, from the Canadian provinces south to North Carolina. All *Coptis* species are similar in appearance and equally useful. Check with your local herbarium or a plant reference specific to your region for the species native to your bioregion.

Applications: The primary active ingredients in goldthread are berberine and coptine, bitter alkaloids with strong antibacterial qualities. The active principles of this plant are similar to those of goldenseal, a plant whose survival has been seriously compromised by market pressure.

For antibacterial purposes, goldthread is an excellent substitute for goldenseal in situations where pathogenic microbes can be directly attacked, such as bacterial infections of the throat and mouth or in the treatment of con-

Goldthread   *Coptis occidentalis*

junctivitis. Goldthread was once a popular treatment for canker sores, which explains its common name "canker root." In recent laboratory studies of *C. sinensis,* a species indigenous to China, goldthread demonstrated promising activities against HIV, infectious hepatitis, and various strains of influenza virus.

Alternatives and Adjuncts: If you are searching for an herbal antimicrobial, there is no reason to harvest goldthread. Oregon grape or barberry (*Berberis* species), another genus of plants rich in berberine, are abundantly available in most of *Coptis's* range. The *Berberis* clan lives in habitats far less sensitive to the consequences of human activities, offers more medicinal material on a per plant basis, and currently is under no threat of being wiped out.

Propagation and Growth Characteristics: Goldthread is a perennial shade plant that requires abundant rich, undisturbed organic matter. It does not transplant well, probably because of its delicate, stringy, small rhizomes. The seeds are small and difficult to collect—you have to anticipate the period immediately before the mature fruits open, or the seeds are lost. Immature seeds generally are not viable.

Goldthread is very hardy; even beneath a deep blanket of snow it is usually in good condition.

Gathering Season and General Guidelines: If you harvest this plant (perhaps in an urgent field situation), take it from a relatively dry area. The rhizomes generally spread through the mat of forest debris, not the soil beneath, which makes them highly vulnerable to damage. Gather from the edges of stands along a well-established foot trail.

Grasp the plant at its base and gently pull up until no more than 3 inches of rhizome is exposed; clip it off with a sharp knife or pruning shears. Where the forest compost is particularly thick and spongy, a gentle pull will do—the roots will come up like threads pulled through lawn clippings.

Care after Gathering: Chew the fresh, dried root or make it into a mouthwash (infusion) for treating mouth ulceration, bacterial infections, canker sores, and so on.

Plant-Animal Interdependence: Like wild ginger *(Asarum caudatum),* goldthread adapts well to full shade where other plants generally do not grow. In dark, forested regions of the northern Rockies, it is typically the predominant plant of the forest floor. The rhizomatous root system promotes soil aeration and water infiltration. The plant breaks up an otherwise impervious mat of forest debris, allowing access for insects and microorganisms critical to the production of compost and the release of plant nutrients. This paves the way for companion plants that could not otherwise adapt. Many of those other plants serve the needs of the pollinators and other organisms important in the biocommunity.

Those delicate, complex biocommunities where *Coptis* lives rely heavily on natural balances maintained through an uninterrupted interdependency between all involved organisms—if one element is removed, the collective whole may fail. In areas denuded of plants like goldthread, you're likely to find a blank or at best sparsely vegetated forest floor.

Tread Lightly: Goldthread is not a viable herb of commerce. It lives in a sensitive environment that is rapidly disappearing under the combined pressures of logging, grazing, and recreational activities. The plant is tiny, and its root yields a disproportionately small amount of medicine for the price of killing the host plant. This is a plant to worry about and protect, not to exploit. Goldthread offers us a chance to redefine what "value" really means and to take the gift of healing to heart instead of to the bank. Instead of harvesting goldthread from the wild, buy *Coptis sinensis*, an Asian cultivar, from your local herb retailer.

WARNING! This plant should not be used during pregnancy.

Gravel Root

Eupatorium purpureum

Sunflower Family

Compositae

Other Names: Joe-Pye weed, sweet Joe-Pye, spotted Joe-Pye, queen of the meadow, kidneywort, purple boneset, trumpet weed, gravel weed

Parts Used: The rhizomes and leaves

Actions: Diuretic, anti-inflammatory, tonic, antilithic, and antirheumatic

Gravel root is a large perennial, sometimes reaching 12 feet tall. The lanceolate leaves have short leaf stems (petioles) and grow directly off the main stem in whorls of four or more. The leaves are distinctively textured and have coarsely serrated margins. The stems are smooth, succulent, and covered with purple spots, especially at the leaf axils. The flowers are small and pink, in showy, slightly globe-shaped clusters at the top of the plant. Spotted gravel root *(E. maculatum)* shares habitat with and closely resembles *E. purpureum,* except the mature plants tend to be much smaller. As medicinal herbs, the two are equally useful.

Boneset *(E. perfoliatum),* a close relative of gravel root, is proportionately smaller and offers a different range of therapeutic usefulness.

Habitat and Range: Gravel root is common at the edges of moist forest clearings, where it often forms dense thickets. Its range includes most of the eastern United States, from Minnesota and Nebraska east to the Atlantic and south to Florida. In the West, *E. maculatum* grows sporadically from southern British Columbia to New Mexico.

Applications: This herb is traditionally used for expelling small kidney and bladder stones (gravel) from the urinary tract. It is considered especially useful for reducing the pain and inflammation of cystitis and is sometimes used to treat a swollen prostate. The diuretic activities of gravel root may help in the elimination of excess uric acid from the body, making it potentially useful in the treatment of gout and other forms of rheumatism.

Alternatives and Adjuncts: For removal of gravel from the urinary tract, gravel root is often combined with demulcent herbs, such as marshmallow root *(Althea officinalis),* fireweed, plantain, or corn silk *(Zea mays).* In urinary disorders involving inflammation or bleeding, gravel root is sometimes combined with stronger astringents, such as uva-ursi, pyrola, pipsissewa, alum-root, or cranesbill geranium.

Propagation and Growth Characteristics: This perennial is easy to grow. It sends up shoots from rhizomes soon after it is established and begins forming its own self-sufficient thicket. All it needs is plenty of sun, rich soil, and frequent watering. Plants will do best if sheltered from wind, as the stems tend to snap easily. The seeds are available through specialty catalogs (see "Resource Guide").

Gathering Season and General Guidelines: The leaves are useful anytime, provided they are in healthy condition. Pluck them off, cut them up, and use them fresh in a strong tea. For stronger medicine, use the roots. Gather roots in fall, after the plant has bloomed and the leaves have started dying off, when the useful constituents are most concentrated in the root. When digging the root, always replant part of the rhizome and mulch it with plenty of grass, dried leaves, or whatever is at hand.

Care after Gathering: The roots can be dried or used fresh. To dry the roots, cut them into 1-inch pieces and spread them on a clean sheet of butcher paper in an area with good air circulation. Rearrange them frequently to prevent mold. If properly dried and stored, they will keep for at least a year or two. The root is best used as a decoction (a simmered tea).

Gravel Root   *Eupatorium purpureum*   Gravel Root (young)   *Eupatorium purpureum*

Plant-Animal Interdependence: A thicket of gravel root is prime habitat for birds, small mammals, amphibians, reptiles, and insects. Where stands of gravel root line the edges of woodland clearings, they are a sanctuary for creatures who use or must traverse the clearings in their daily endeavors. The big, showy, fragrant flower clusters are a critical source of pollen for resident and migratory insects. The juicy stems and nutrient-rich leaves break down into copious amounts of compost each year.

Tread Lightly: Gravel root is abundant and it coexists with humanity quite well. Despite its abundance, however, we mustn't overlook its ecological importance, especially in areas where it stands in thick proliferation. Take a careful look under some of the leaves to glimpse the menagerie of creatures who reside beneath gravel root, and if an animal is using the patch, move on to another.

Hawthorn

Crataegus species

Rose Family
Rosaceae

Other Names: White thorn, mayblossom
Parts Used: In early spring, the flowering end-limbs (leaves, flower buds, thorns, and all); the berries in fall
Actions: Vasodilator, hypotensive, tonic, and nutritive

Hawthorn is a small deciduous tree or large shrub (up to 16 feet tall) easily recognized and quickly remembered by its nasty 1- to 3-inch curved thorns strategically spaced along the branches—often at eye-level. The alternate leaves, narrowly fan-shaped or ovate, have short leaf stems (petioles). The margins of the 1- to 2-inch-long leaves are toothed, with the teeth all pointing distinctly forward. The white, ¼-inch flowers form flat, terminate clusters; each blossom has five petals and many stamens. In full bloom, the blossoms may have an unpleasant "dead" odor. In late summer, the flowers are replaced with clusters of red to black berries, each containing two to five seeds. *C. oxyacantha*

Hawthorn   *Crataegus douglasii*

and *C. monogyna* are the primary hawthorns of commerce, and *C. douglasii* (black hawthorn) is one of the most common and widespread wild hawthorn species of North America. Very little study has documented which hawthorn species is most useful, but herbalists agree they all hold therapeutic value.

Habitat and Range: The *Crataegus* genus is large and varied, with hundreds of species in North America (all of which readily hybridize). Most species grow in riparian thickets.

Applications: Hawthorn (*Crataegus* species) is a classic example of an herb that is truly tonic—it does not stimulate an immediate change in the body but provides specialized, long-term support to the cardiovascular system as no other food, herbal medicine, or drug can. Herbalists have regarded hawthorn berries as one of nature's best and safest heart and vascular tonics for thousands of years. Extensive scientific study validates hawthorn's usefulness in this capacity, and herbalists and researchers agree that hawthorn benefits the heart and arteries in at least three ways:

1. Hawthorn dilates coronary vessels and vessels of the brain, helping to increase circulation and the transport of nutrients and oxygen through the body. It accomplishes this in an effective and unique fashion: as it dilates major vessels it also increases blood flow from the heart to compensate for any reduced arterial blood volume. In other words, it helps the body push more blood around by increasing cardiac output and decreasing blood flow resistance in the arteries—more blood flow at less pressure.

2. Hawthorn possesses antioxidant properties; it scavenges free-radicals, which are known to rob the blood of oxygen and may lead to various vascular diseases.

3. Hawthorn steadies and strengthens a weak or erratic heartbeat. In fact, it has been shown to act as a possible alternative to the drug digitalis, and may serve as a potentiating adjunct to this cardiac drug.

All those activities are largely because of the vast array of flavonoid constituents in hawthorn. Flavonoids are red pigments found in many fruits and vegetables, and hundreds of studies indicate that these compounds are essential in maintaining disease resistance and the integrity of smooth muscle tissues in the body. Hawthorn may be the richest natural source of these vital nutrients.

Alternatives and Adjuncts: For tonic circulatory system support, herbalists often combine hawthorn with garlic, ginkgo *(Ginkgo biloba)*, yarrow, cayenne *(Capsicum minimum)*, culinary ginger *(Zingibur officinale)*, or horse chestnut *(Aesculus hippocastanum)*. For high blood pressure, they combine it with garlic or yarrow.

Propagation and Growth Characteristics: Many hawthorn species are available through nurseries. They are not difficult to grow—their main requirement is ample water—but they grow slowly. To propagate this shrub, purchase one at least three years old. It tends to grow outward as much as upward, so plant it where it has room to stretch and brandish its nasty thorns!

Harvest Season and General Guidelines: Herbalists and researchers argue about which parts of this shrub make the most effective medicines—the flowering spring branches or the fall-harvested ripe fruit. I combine them for maximum effectiveness.

Gather from the outermost, small limbs when they are flowering and the leaves are just starting to bud out, usually in early spring. With sharp pruning shears, clip 8 to 12 inches from the tips of healthy limbs no thicker than a pencil that bear plentiful leaves and flowers. Gather only from limbs within easy reach and never so much that you create a noticeable difference in the tree's overall appearance.

In fall, gather the berries when they are fully ripe; they will be crimson red or dark purple (depending on species) and slightly sweet. Don't use a ladder—gather conservatively from limbs within easy reach.

Care after Gathering: Tincture the flowering twigs right away. The berries, too, are best if tinctured fresh, but you can dry them for teas if you wish. When you have tinctures both of the spring-harvested limbs and the fall-harvested berries, combine the two to make a synergistically complete formula.

∽Tincture

Flowering twigs, cut as fine as possible with shears: 1:2 ratio in 60 percent alcohol. *Fresh, crushed berries:* 1:2 ratio in 60 percent alcohol. *Dried berries:* 1:4 ratio in 60 percent alcohol.

Plant-Animal Interdependence: Hawthorn is a critical source of food and nesting habitat. Everything from moose to birds relish the berries, and the dense stature and thorny attitude of the shrub furnish a safe place to build a nest.

Tread Lightly: Because this shrub or small tree is so important to wildlife, the ecoherbalist must be especially careful not to disrupt natural balances in and around a stand of these plants. Disperse your harvest over several locations. This plant offers a wealth of information about how a tightly interconnected ecosystem lives from day to day. When you visit a stand of hawthorns, pause and learn something new about how life around you coexists.

Horehound
Marrubium vulgare

Mint Family
Labiatae

> **Other Names:** White horehound, marrubio
> **Parts Used:** The leaves primarily, and the other aboveground parts
> **Actions:** Expectorant, bitter, and diaphoretic

Unlike many members of the Mint family, horehound has little or no discernible odor. Like other mints, horehound has four-sided (square) stems, opposite leaves, and flower clusters at the leaf axils of the upper plant. Horehound stems and leaves are distinctively woolly. The leaves are ovate, ½ to 1½ inches wide, wrinkled, and coarsely toothed. The flowers are tiny, white, and tubular and form in whorled clusters that develop into bunches of small, spiny burrs. Horehound typically becomes densely branched when mature and can grow to 3 feet tall, but usually it grows in short clumps.

Habitat and Range: Horehound is a Eurasian import that has found its way into dry, disturbed areas across most of the Northern Hemisphere. It is a common inhabitant of vacant lots and roadsides.

Applications: Time-proven as an expectorant, horehound is in many commercial brands of cough drops, traditionally known as "horehound candy." I believe that name indicates two things: (1) how a foul flavor can be improved by adding copious amounts of sugar, and (2) how far we will go to get a cough remedy into a child. Horehound tastes awful . . . not just bitter but horrible!

Nevertheless, this herb is an excellent expectorant, and it pains my taste buds to acknowledge that it does its job best when used in the purest form possible, such as in tea, tincture, or a honey-based syrup. Horehound is indicated when a tickling cough becomes redundant and the hacking isn't producing anything. Horehound is also an excellent bitter (see the uses described for dandelion).

Alternatives and Adjuncts: In homemade cough syrups, horehound combines well with false Solomon's seal, arrowleaf balsamroot, marshmallow *(Althea officinalis),* or plantain. Mullein, coltsfoot (*Tussilago* or *Petasites* species), or grindelia *(Grindelia squarrosa)* are all worthy as substitutes. For antiviral and immune modulating activities, try adding arrowleaf balsamroot, echinacea, or lomatium to horehound.

Propagation and Growth Characteristics: Horehound is a deep-rooted perennial that isn't picky about soil quality. It likes hard, dry gravel or clay low in organic matter. Though the plant will flourish in deep, rich soils, the best horehound is found in poor soil.

It is an easy plant to establish from seed or transplants and is an excellent addition to the herb garden. A word of advice from my personal experience, though—it may march its way out of the garden!

Gathering Season and General Guidelines: Cut the stems or harvest the leaves as you need them, leaving plenty of flowers so the plants can reproduce. Although you can use horehound at any stage of growth, it tends to be most potent during hot, dry weather and when the plants are in full bloom.

Care after Gathering: Use fresh for a foul-tasting but effective tea or cough syrup.

∿Tincture

Fresh plant: 1:5 ratio in 50 percent (100 proof) alcohol. To make a syrup, see false Solomon's seal.

Plant-Animal Interdependence: Horehound typically grows as an earth regenerator, and once established it usually stays in its chosen biocommunity forever. This plant provides a dependable source of cover for small animals and insects in areas where other cover may be scarce, such as the edges of dirt roads.

Horehound is an effective pollinator attractor and contributes at least small amounts of organic matter to the depleted soils where it resides. Its deep root system is effective in preventing erosion.

Tread Lightly: As with all imported plants, take care to avoid introducing horehound into an area where it does not already exist; inspect your clothing for hitchhiking seeds. Assess the plant's roles in soil conservation and also the quality of insect and small animal cover before gathering.

Horehound *Marrubium vulgare*

Horsetail

Horsetail Family

Equisetum arvense and *E. hyemale*

Equisetaceae

Other Names: Scouring rush, jointed grass, puzzle grass, pewterwort
Parts Used: All aboveground parts of the young, green, sterile growth
Actions: Diuretic, tonic, and hemostatic

The Horsetail family divides into three groups: annual species with separate, distinctly different fertile and sterile stems; annuals with sterile and fertile stems that are similar in appearance; and evergreen perennials with fertile and sterile stems that look alike. Despite variations, the entire family shares fundamentally similar characteristics: hollow, distinctively grooved and jointed stems, and leaves that are scalelike and dark, resembling sheaths surrounding the stems at the joints. The plants reproduce by spores from a terminal cone on the fertile plants. Two common and widespread species are *E. arvense* (common horsetail) from the first group, and *E. hyemale* (common scouring rush) from the third group. In early spring, *E. arvense* produces a small (3- to 12-inch) fertile stem that lacks chlorophyll. It dies back as its larger, green, sterile counterpart matures. The subsequent sterile stems (6 inches to 2 feet tall) have whorled branches that give the overall appearance of a green bottle-brush. *E. hyemale* is much larger (up to 5 feet tall) and lacks any branching characteristics. Its looks like a prehistoric cross between miniature bamboo and an asparagus spear. The two species commonly grow side-by-side.

Habitat and Range: Horsetail frequents moist meadows, wooded areas, stream banks, and lakeshores from sea level to about 10,000 feet. Some species grow directly out of shallow water in mountain lakes. Most horsetails prefer at least partial shade and deep, rich soil. Horsetail is common across North America and in most of the Northern Hemisphere.

Applications: Horsetail contains an impressive amount of silicon, a chemical element critical in the growth and regeneration of bone and connective tissues. Silicon is one of the most abundant elements in nature. The problem with most sources of silicon, though, is that the body cannot absorb them, and compounding the problem, the body does not produce all the silicon it needs. Therefore, it must derive this critical element from outside sources. Horsetail provides the body with a supplemental source of silicon that the body can assimilate on an "as needed" basis. For this reason, horsetail is regarded as one of the best herbs for treating bone fractures and tendon and ligament injuries.

Some interesting studies suggest that aluminum-silicon balances in the body may be related to senility. When the balance between these two elements favors aluminum, senility may progress faster. Horsetail supplements, then, might help slow the process.

Horsetail is an excellent hemostatic, and herbalists often use it to stop internal or external bleeding. It is also mildly diuretic, and can be a useful urinary tract tonic for ailments ranging from cystitis to incontinence.

Alternatives and Adjuncts: To add an antibacterial element to augment its use for urinary tract infections, horsetail combines well with echinacea, arrowleaf balsamroot, or Oregon grape. Marshmallow root *(Althea officinalis),* or a good tincture of plantain, will add a soothing, mucilaginous quality. As a tonic bone and connective tissue regenerator, horsetail combines well with nettle.

Propagation and Growth Characteristics: Horsetail usually grows as an annual, with some species (such as *E. hyemale*) growing mainly as evergreens.

Horsetail *Equisetum arvense*   Horsetail *Equisetum hyemale*

In areas where climatic conditions are severe or human interference is consistently high, evergreen horsetails may grow as annuals.

Although it's possible to cultivate horsetail, attempts at making this a garden herb are usually futile. The rhizomes generally do not transplant well, and the nearly microscopic spores must be collected at their moment of optimum fertility for immediate distribution. Although the germination rates of the spores are generally low, I would opt for this method of propagation. Distribute spores beneath a dripping spigot or in a moist drainage area.

Ironically, if horsetail moves into your garden on its own, it can be difficult to eradicate. Its voluntary presence in the garden usually indicates such problems as high groundwater and/or poor soil drainage.

Gathering Season and General Guidelines: Gather the sterile (green) stems while their little branches are still pointing upward (generally in midspring). In older plants the silicon constituents crystallize with other compounds as the plants age, making them less soluble in water and the plant less useful.

Gather from the periphery of dense, healthy stands, taking care not to compact any more soil than absolutely necessary to access the plants. Clip the stems with sharp pruning clippers an inch or more above ground level.

Do not gather horsetail from heavily fertilized areas or areas with high irrigation runoff, as this plant readily stores nitrates, selenium, and a variety of heavy metals. For the same reason, be wary of livestock grazing areas, which are often subjected to herbicides.

Care after Gathering: Because this plant lives in or near water, the fresh plant will not readily infuse into water; it must be dried or tinctured before it is useful internally. Like all herbs gathered from a moist environment, horsetail needs ample air circulation and dry conditions to prevent mold growth. Small bunches (1 inch or less) can be tied and hung in an airy, dark location, or spread the plants on a nonmetallic surface and stir them frequently.

∾Tincture

Fresh, chopped herb: 1:2 ratio in at least 50 percent alcohol. *Dried herb:* 1:5 ratio in 50 percent alcohol. Unless the tinctures are filtered through paper or fine cloth, the gritty particles contained in the fluid can irritate the urinary tract and kidneys if used over prolonged periods.

Plant-Animal Interdependence: This rhizomatous plant benefits the soil structure of wet environments, particularly near moving water, where the plant's roots help prevent soil erosion.

A stand of horsetail is excellent shelter for birds, reptiles, and amphibians. Horsetail is one of the oldest inhabitants of our planet—only the size has changed during the past 300 million years! Today, deer, moose, and the descendants of dinosaurs we now know as birds forage horsetails.

Tread Lightly: The soil where horsetail grows is usually vulnerable to the harmful effects of human feet. Whenever possible, gather from the margins of the driest stands you can find and avoid those that are among a profusion of other plants.

WARNING! Unless your tinctures are filtered through paper or fine cloth, the gritty particles contained in the fluid can irritate the urinary tract and kidneys if used over prolonged periods.

Hound's Tongue
Cynoglossum officinale

Borage Family
Boraginaceae

Other Names: Dog burr, dog's tongue, gypsy flower, sheep lice, woolmat
Parts Used: The entire flowering plant, including the root
Actions: Vulnerary, expectorant, demulcent, and diuretic

Hound's tongue is a member of the Borage family and looks like its close relative, comfrey. The leaves are opposite, lance-shaped, rough and hairy, 1 to 3 inches wide, and up to 12 inches long. Hound's tongue usually grows as a biennial, producing only a basal rosette of leaves the first year and a flowering stalk that may reach as tall as 4 feet during its second and final year. The root is long and tapered, like a carrot.

The terminate flowers are reddish purple and form a loosely arranged bouquet at the top of the mature plant. The flowers develop into flat, tongue-shaped burrs that have a remarkable capacity for sticking to *anything*. The seeds typically remain on the dead stalk for a long time, turning light gray and making late-season identification easy.

Habitat and Range: Hound's tongue is a European import that has made itself at home in pastures, ditches, and at roadsides across North America. Although it has a reputation for being toxic to livestock, it is curiously common in areas that have been disturbed by free-ranging cattle—where the cows appear to be just fine!

Hound's tongue is not particular about its habitat but is most abundant in areas where the seeds can cling to passersby. It is common on the margins of trails and roadways, from sea level to about 9,000 feet.

Applications: Hound's tongue is a good alternative to comfrey. Like its relative, it contains allantoin and the alkaloid heliosupine. Allantoin speeds the process of granular cell reproduction at injury sites. In topical applications, hound's tongue oil or salve help accelerate the healing of skin irritations, minor burns, cuts, scrapes, and, particularly, insect bites. Although herbalists sometimes use this plant internally as a sore throat and cough remedy, exercise care and moderation because, like comfrey, hound's tongue contains potentially carcinogenic alkaloids. I do not use this herb for internal applications.

Alternatives and Adjuncts: For topical use in first-aid applications, hound's tongue should be combined with an antibacterial, such as Oregon grape, usnea lichen, or bee balm. For internal expectorant or demulcent applications, herbs such as false Solomon's seal, arrowleaf balsamroot, plantain, mullein, or grindelia may be safer than hound's tongue.

Propagation and Growth Characteristics: No secrets here! Hound's tongue is a highly successful weed in North America. It is easy to start from seed, and if introduced into the herb garden it can take over. The plant's seed distribution efficiency becomes obvious once you have wrestled the burrs from your clothing or Fido's fur. Hound's tongue is a biennial that may grow as a perennial depending on elevation and climate.

Gathering Season and General Guidelines: Gather this plant when it is just beginning to bloom, generally in late spring or early summer. Dig the roots with a narrow, sharp shovel, such as a ditch spade, taking care not to disturb any of its less durable neighbors.

If you end up gathering roots after the plant has gone to seed, carefully remove any hitchhiking burrs from your clothing, hair, or dog before moving on. This plant is commonly sprayed with herbicides. Check with your local

Hound's Tongue   *Cynoglossum officinale*   Hound's Tongue fruits   *Cynoglossum officinale*

extension office for "weed abatement" programs in your area before you begin gathering, and always keep in mind that introducing this plant into new areas can lead to the introduction of chemicals as well.

Don't gather this herb within 100 feet of any roadway, as it accumulates lead compounds and other toxic residues on its hairy leaves.

Care after Gathering: Dry the upper plant and the chopped root using standard techniques (see "Harvesting and Handling Herbs in the Fields"). The dried herb will store in quality condition for about six months.

To make an oil infusion, place the chopped dried herb in a glass jar. Cover the material with enough olive oil to leave a ½-inch film of clear oil above the herb. If air contacts the herb, the infusion process will spoil. Cover the jar with a tight-fitting lid and let it sit for at least one month. Then strain and press the mixture through unbleached muslin or a coffee filter and store the oil in the refrigerator.

Plant-Animal Interdependence: The long taproot benefits high-impact areas where erosion may be a problem. The sturdy, persistent root can penetrate hardpan soils, promoting aeration and the introduction of less adaptable plant companions.

The flowers of hound's tongue are effective pollinator attractors, and the leaves provide shelter and habitat for beneficial insects and small animals. Where hound's tongue grows at roadsides, it provides refuge for insects and small animals scurrying out of harm's way.

Tread Lightly: Hound's tongue is an abundant plant that seems to be good at resisting the pressures of human interference. It is also the target of herbicides. Please leave the seed distribution of this plant to our animal allies, and whenever possible, discourage the use of herbicides in natural environments. Take extra care to avoid inadvertently carrying the burrs to areas it has not invaded.

WARNING! This plant contains alkaloids that may be carcinogenic or damaging to the liver if ingested in large quantities.

Juniper
Juniperis species

Cypress Family
Cupressaceae

Other Names: "Cedar," creeping juniper, prickly juniper, creeping savin, common juniper

Parts Used: The berries, leaves, and branches

Actions: Astringent, diuretic, bacteriostatic, tonic, and hypoglycemic

Several juniper species inhabit North America. Generally, the various species of this large genus are differentiated by size (trees or low-growing shrubs) and the specific characteristics of their leaves (needlelike or scalelike). The most widespread species is common juniper *(J. communis),* a ground-hugging shrub with sharp, ¼- to ½-inch "needles" arranged in whorls of three. Of the tree junipers, western juniper *(J. occcidentalis),* Rocky Mountain juniper *(J. scopulorum),* and Utah juniper *(J. osteosperma)* are common to their respective habitats in the western United States and typically are the predominant foliage of the landscape. The leaves of these species are the scalelike variety. All junipers can be easily identified by their foliage and cones (commonly referred to as "juniper berries"), which have the distinctive aroma of gin. The conspicuous ¼- to ⅜-inch female "berries" are dusty blue, and grow from leaf axils (where leaves join the branches), where they may remain for two to three years before ripening and falling to the ground. The small, male cones, borne alone at the tips of the branches, are inconspicuous.

Habitat and Range: Common juniper is typically a mountain shrub, growing on rocky hillsides and forest clearings up to about 10,000 feet. It is widespread from Alaska to California and across much of temperate North America. The tree species are high-desert dwellers of eastern California, Nevada, Utah, Colorado, Wyoming, Arizona, and New Mexico. Several cultivated species of juniper are landscape shrubs, worldwide.

Applications: Herbalists regard juniper as a strong urinary tract tonic and disinfectant. The berries, and sometimes the leaves, are used to treat cystitis, urinary tract bleeding, and various types of kidney disease. Because of juniper's strong volatile oil and tannin constituents, however, long-term internal use of this plant may have a reverse, irritating effect on the urinary tract and kidneys.

A decoction of the leaves and the berries is sometimes used in the remedial treatment of hemorrhoids and inflammations of the skin, and to reduce postpartum swelling. The strong, astringent actions of juniper make it useful in these applications, as the tannin constituents have a pronounced tightening effect on swollen tissues.

Recent studies indicate that a decoction of juniper berries may be useful in the treatment of insulin-dependent diabetes; juniper lowers blood glucose levels by increasing glucose uptake in the diaphragm while increasing the

Common Juniper   *Juniperis communis*

Utah Juniper   *Juniperis osteosperma*

release of insulin from the pancreas. The result: a lower mortality rate in people with insulin dependent diabetes and new hope for those with other forms of diabetes. In a study conducted in Taiwan, a compound derivative of the berries—14 acetoxycedrol—had anticoagulant effects in the blood and a relaxing action on vascular tissues, actions that may help reduce the risk of cardiovascular disease.

Juniper berries also relieve pain in muscles and joints and ease the symptoms of arthritis and rheumatism when used topically.

Alternatives and Adjuncts: Where a topical or urinary astringent is indicated, alumroot, uva-ursi, pipsissewa, or pyrola are good substitutes.

Propagation and Growth Characteristics: Juniper is a drought-tolerant evergreen hardy enough to survive even the harsh winters of the Canadian interior. It tolerates any soil, so long as ample stabilizers, like rocks, can anchor its roots.

Juniper tends to be slow when started from seed and does not transplant well from a natural environment. You can purchase young shrubs through a commercial nursery for introducing into the herb garden. The berries of this plant require at least two years to reach maturity.

Gathering Season and General Guidelines: Gather juniper berries when they are dark blue and juicy. The leaves and branches can be gathered anytime. Be conservative, though—this plant is an important source of food and shelter for many animals, and it regenerates very slowly.

Care after Gathering: Use the leaves and branches fresh, for teas or decoctions. Use the fresh berries medicinally or for cooking, or dry and store them for several years. For tincture, I like a combination of the leaves and the fresh berries, although most herbalists are partial to the berries alone.

∾Tincture
Fresh leaves and/or berries: 1:5 ratio in 75 percent alcohol. *Dried berries:* 1:5 ratio in 50 percent alcohol.

Plant-Animal Interdependence: Juniper is an important food source for deer, elk, moose, bear, and rodents, and is especially relished by birds.

At my home, in western Montana, gathering juniper berries from the backyard bush depends on the demands of the resident spruce grouse and the mobs of migrating grosbeaks and sparrows that pass through each year. The squirrel who has made its home directly above the patch has first choice of what's left, and then I get in line behind a collaboration of deer, elk, and ground-dwelling rodents. Needless to say, I get only a few berries each year!

The low-growing, densely branched shrubs provide good cover and habitat for a wide variety of organisms. Where juniper grows on steep hillsides, it helps prevent erosion and holds back nutrient-rich forest debris that might wash downslope during heavy rains. In high alpine areas, junipers typically grow directly out of rock fissures, with a collection of dependent flora growing from the alluvial accumulation at the bases of their tough trunks.

Tread Lightly: The amount of damage our juniper harvesting causes is inversely proportionate to our level of awareness about the microecosystem where it grows. Juniper is an important food and shelter source for many sometimes transient organisms. Migratory birds frequent the shrub near my house, but I acquired that knowledge over years, and even a high degree of familiarity cannot foresee a drought, a fire, or an increase next year in the deer population.

Try to become intimately familiar with the daily activities of the juniper stands where you harvest. Then gather carefully and thoughtfully, keeping records of any new observations you make during your visits.

> **WARNING! Juniper can irritate the urinary tract and kidneys if used excessively. Do not use during pregnancy or in the presence of kidney disease without the guidance of a qualified health care practitioner.**

Lomatium
Lomatium dissectum

Parsley Family
Umbelliferae

Other Names: Biscuitroot, big medicine, bear medicine, consumption root, western wild parsley, mountain parsley, desert parsley, fern-leafed lomatium

Parts Used: The root

Actions: Antiviral, expectorant, antimicrobial, and immunostimulant

With few exceptions, *L. dissectum* is a typical member of the Parsley family. Flowers form in umbrella-shaped terminal inflorescences, which range from white and yellow to pink and purple. The stems are hollow and the pinnately divided, compound leaves resemble parsley (depending on species). More than seventy species of *Lomatium* inhabit the western mountains of the United States and Canada. The species frequently hybridize, making identification difficult and sometimes frightening, as several species of this plant share a striking resemblance to poisonous hemlock parsley *(Conium maculatum)*. Unlike hemlock parsley and most other Umbelliferaes, though, lomatium grows in dry, rocky, often steep habitats. Hemlock parsley and other toxic umbels generally require consistently moist, deep soils. Lomatium *can* grow in moist soils and sometimes shares the habitat of its poisonous cousin, so check your local plant key for species in your area. Apply this simple rule for identifying the *Lomatium* genus by habitat: if the soil is dry and mainly void of organic matter, it's *probably* a *Lomatium*. But if the plant is growing in moist soil, it could well be a poisonous plant. Once you've determined habitat characteristics, look for distinctive physical characteristics, such as the large (½- to ¾-inch), oval seeds of many species. *L. dissectum* is the largest member of the genus, the one commonly recognized as a medicinal plant. This species may grow to 40 inches tall and can be readily differentiated from smaller *Lomatium*s (such as *L. cous*) by its size. All species of *Lomatium* have edible roots and leaves; *L. cous* is the best tasting.

Habitat and Range: *L. dissectum* typically grows in dry, rocky soils. Many species are common to rock slides and steep hillsides, sometimes growing directly from fissures in vertical escarpments. It ranges south from British

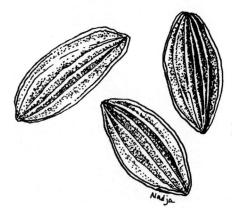

Lomatiun dissectum seeds
(3 times actual size)

Columbia throughout the Rockies to Arizona and west into the mountains of California.

Applications: Blackfeet and other Indians used lomatium extensively as a food and medicine. American physicians and pharmacists routinely prescribed it as an antiviral before the advent of vaccines and allopathic remedies. Recent studies indicate that lomatium may inhibit various strains of influenza virus, especially those that infect the respiratory tract.

Lomatium   *Lomatium dissectum*

Because of a frightening, worldwide emergence of newly discovered viruses, lomatium is a prime candidate for commercial exploitation. As this plant's popularity grows, its use in applications that exceed its therapeutic capacity also grows. Reserve lomatium for cases of flu and other serious viral infections that originate in the respiratory tract. Lomatium has antimicrobial, immune-stimulant, and expectorant qualities, but other herbs work just as well in these capacities. For symptoms of a common cold, or to boost the immune system, choose a plant that is more abundant, more widespread, and more resilient to human impact.

Alternatives and Adjuncts: For applications where an immune stimulant is indicated, look to arrowleaf balsamroot and echinacea as better options than lomatium. Where an expectorant is indicated for systemic support during a common cold, investigate using mullein leaf, coltsfoot (*Petasites* or *Tussalago* species), grindelia, or goldenrod. For respiratory infections of bacterial origin, choose bee balm, yarrow, or arrowleaf balsamroot over lomatium.

Propagation and Growth Characteristics: If you want to grow lomatium, my best wishes are with you! Lomatium is a slow-growing perennial that is difficult to propagate. The plant increases in size through the production of root crowns but distributes its population only by seed. Each mature plant produces hundreds of seeds, which are dispersed mainly by gravity or runoff, with a low germination rate.

The plants do not transplant well, as the roots tend to bleed profusely from even the mildest injury. Successful division of root crowns is a poor-odds endeavor. Try starting this herb from seed. Gather a few from each of a number of plants, and stratify them or sow them in the fall, covering the seeds with ¼ inch of soil in a dry, rocky location, preferably a hillside. In my experience, about one in five seeds will germinate and about one in five of seedlings will survive the first year. Give seedlings plenty of water during their first year, then wean them and let nature take over. And after all that, you may die of old age before the plants reach a useful size! I once thought that a mature *L. dissectum* might be fourteen to twenty years old, but after watching seedlings grow by millimeters each year, and after talking with wildcrafters who are well acquainted with this plant, I now think that some of the lomatiums I have killed through harvest may have been more than fifty years old. Before harvesting, ponder this important question: Does my need justify the permanent elimination of a powerful healing ally? Then, if you decide to harvest, be careful to ensure that none of the plant goes to waste.

Gathering Season and General Guidelines: If you choose to dig lomatium, gather the mature roots after the foliage has died back and the flowers have gone to seed, usually in late summer. Use a narrow, round-nosed shovel or a geologist's hand pick to dig the often stubborn roots with a minimum of effort and interference. If any seeds remain on top of the dried stems, plant a few at the exact spot where you dug the root, and scatter the rest in a manner that replicates natural processes as closely as possible.

Wildcrafters seem to agree that lomatium should be harvested from the bottom of hillside stands; the stand is naturally maintained by the distribution of seeds from the upslope plants. This is good thinking, but the wildcrafter needs to consider additional issues before harvesting this plant. What about the succession of wildcrafters who gather from the bottom of the same stand? The bottom of the stand could be forced steadily uphill. What condition are the mother and father plants in? Consider all the variables, then gather conservatively and as infrequently as possible from multiple sites.

Care after Gathering: A large lomatium root will provide the herbalist with a supply of tincture that will last several years. I prefer to tincture the fresh root, but the dried root can be used as well. Brush excess dirt from them (do not wash them or you might be inviting mold) and quarter them lengthwise. Dry the roots in an open paper bag, or, chop fresh roots and use them in tinctures. The dried roots will remain in good quality for six months or more if properly stored in plastic bags, but alcohol tinctures will last forever.

⟶Tincture

Fresh root: 1:2 ratio in 70 percent alcohol. Lomatium does not take to glycerine. *Dried root:* 1:5 ratio in 70 percent alcohol.

Plant-Animal Interdependence: The large taproot of this plant is instrumental in erosion control and often creates a "rock dam" in slide areas. This function becomes obvious once you have experienced a small rock slide while pulling a large root out of the ground.

The yellow or purple flowers are important pollinator attractors, and bears, bighorn sheep, and other large animals sometimes eat both the upper plant and roots.

Tread Lightly: Lomatium is not rare or endangered; it is still profusely abundant in remote areas of Nevada, Idaho, and Montana. But increased consumer demand, habitat vulnerability, and the finicky, slow-growing characteristics of this plant put existing populations at risk for overharvesting. So, why should we label this plant as "a species at risk" when most of its population is centered in areas where nobody wants to pack it out? The current market price does not justify expensive and arduous expeditions into rugged wild areas, but overharvesting from the easily accessible sources will force the price up. The herb industry *will* go after lomatium in pristine, remote habitats if the price is right. Its fate rests in how responsibly it is harvested and used as an herbal medicine. Lomatium should be reserved only for specific cases where a strong respiratory antiviral is indicated.

> **WARNING! This plant closely resembles highly poisonous members of the Parsley family. If you are not absolutely confident that you can differentiate this plant from its look-alikes, don't mess with it!**

Mullein
Verbascum thapsus

Figwort Family
Scrophulariaceae

Other Names: Velvet plant, candle leaf, Indian tobacco, blanket leaf
Parts Used: Leaves, flowers, or roots
Actions: Expectorant, antispasmodic, vulnerary, demulcent, diuretic, antimicrobial, tonic, and insecticidal

In its first year of growth, this conspicuous biennial forms a basal rosette of large (up to 12-inch), broadly lance-shaped, profusely fuzzy leaves. During its second and final year, mullein heads skyward with a stout, central stalk that may be more than 6 feet tall. The numerous yellow flowers then form in a terminate, coblike inflorescence.

Habitat and Range: Mullein is a Eurasian import that has made itself at home in a variety of disturbed sites across North America. It is common to clearcuts, burns, and partially developed lands.

Common Mullein *Verbascum thapsus*

Common Mullein flower *Verbascum thapsus*

Applications: Herbalists use mullein leaves for an expectorant and respiratory antispasmodic. Tea made from it may be especially useful at the onset of respiratory infection, when the upper respiratory tract is acutely inflamed, a tickly spastic cough has ensued, and the subsequent secretion of soothing, cleansing mucus seems long overdue. Some herbalists use this herb in a similar fashion at the onset of asthma attacks, but I have no experience with the use of mullein in this capacity.

Chemical constituents in the flowers are known to be highly antimicrobial, while others (mainly rotenone) are known to have insecticidal properties. Many herbalists use an oil infusion of the fresh flower buds as a treatment for bacterial infections of the ear. Brew the flowers into a tea for topical use against fleas or mites, or use the cooled tea in the garden as an environmentally safe, wide-spectrum insecticide. A word of caution though: rotenone is highly toxic to aquatic life-forms.

The fresh root tincture is used as a specific treatment for urinary incontinence. In this capacity, mullein root may have a tonic action on urinary muscle tissues responsible for urine retention in the bladder.

Alternatives and Adjuncts: Various herbs are useful for different kinds and qualities of coughs. Coltsfoot will usually substitute for mullein leaf, but read up on grindelia, goldenrod, arrowleaf balsamroot, and horehound as other possible adjuncts or alternatives. For antimicrobial applications, combine mullein flower with oil infusions of fresh garlic or Oregon grape root. For urinary incontinence, try nettle root or sweet sumac *(Rhus aromatica)*. Corn silk or marshmallow is often added as a soothing, tonic adjunct in urinary formulations.

Propagation and Growth Characteristics: Mullein is a highly efficient biennial. After the flowers have matured and begun to dry, each forms a capsule that contains a multitude of tiny seeds. Once the capsules begin to open, thousands of seeds are released and distributed by wind, precipitation, or the slightest disturbance of a passerby. The seeds may lie dormant for several years before germination.

Mullein is not particular about soil. It prefers full sun but will tolerate partial shade. It is very cold hardy and drought resistant.

Although this is an easy plant to introduce into the herb garden, I don't recommend it unless you intend to harvest the herb before it matures. Once it goes to seed, it will launch an invasion on your garden.

Gathering Season and General Guidelines: Gather mullein leaves anytime; midseason leaves (May to August) are generally in optimum condition. Gather flowers and/or flowering tops when the buds are about halfway open. Some herbalists tediously pick off each individual flower, instead of cutting off the entire flowering tops. The latter is much easier and produces fine medicines, but the plucked buds produce a prettier oil. It's a matter of how serious you are about Zen and the art of mullein harvest. Dig the root during its first year. Because the plant is biennial, the roots die and dry up after the plant blooms.

Because many people view this plant as a "weed," the wildcrafter should always be suspicious of herbicide use where mullein grows. Do not gather this herb from cultivated areas, where it likely is contaminated with herbicide.

Care after Gathering: Dry the leaves on a nonmetallic surface, or tie them into small bunches (no more than 1 inch thick at the base) and hang them in a dark, airy space. Unlike most dryland plants, mullein seems to be vulnerable to mold while it dries, so watch it closely and rearrange it frequently. The dried leaves will keep for up to six months stored in plastic bags.

Use the flowers and flowering tops for making oils while they are fresh. The active constituents are contained in a dark, tarry substance that you can see by picking apart the flower heads. This substance loses its potency as it dries. Mince the flower heads or buds, place them in a jar, and cover with enough olive oil to leave a ½-inch layer above the herb. The herb must remain submerged in the oil or it will spoil. Seal the jar with a tight-fitting lid, let it sit for about a month, and then strain it through a jelly bag or muslin.

∾Tincture

Fresh leaves or root: 1:2 ratio in 60 percent alcohol. *Dried leaves or root:* 1:5 ratio in 60 percent alcohol.

Plant-Animal Interdependence: Mullein is a loyal earth-regenerator and is one of the first sources of small-animal habitat in damaged areas. Its stout taproot is effective in controlling erosion and breaking up compacted soil in areas where other plants may not yet have recovered from fire or other disturbances; the tall, coblike yellow flowers help reintroduce pollinators.

Tread Lightly: In areas where mullein is busy repairing habitat, leave the plant alone. Gather mullein from areas where it's not performing an urgent task.

WARNING! The seeds and flowers of this plant may be toxic if ingested in large quantities.

Nettle
Urtica species

Nettle Family
Urticaceae

Other Names: Stinging nettle, nettles, ouch!
Parts Used: The entire upper plant; the root for prostate troubles
Actions: Nutritive, astringent, diuretic, tonic, and antihistamine

For people who have learned to avoid its sting and recognize its attributes, nettle is a true delight. But to the unknowing explorer who haphazardly wanders into a stand of these plants, nettle offers a crash course in plant identification (and contact dermatitis!). The plant stems and the underside of the leaves are covered with thousands of tiny, hollow, needlelike hairs that contain a combination of antigenic proteins and formic acid (the latter is also the chosen weaponry of red ants). When contacted with sufficient force, these minihypodermics inject their contents, causing a sudden, burning rash. The pain and tiny blisters usually subside within an hour.

Nettle is an erect plant that may grow as tall as 7 feet. It reproduces mainly from shallow rhizomes and commonly grows in dense colonial patches. The opposite leaves are broadly lance-shaped, with coarsely toothed margins. Flowers form at the leaf axils in inconspicuous, brownish, drooping clusters. The stems are covered with fine, stinging hairs. Young plants typically emerge reddish, turning green as they mature.

Habitat and Range: Several *Urtica* species grow in consistently moist, rich soils, often in roadside ditches or riparian habitats across most of North America. *U. dioica* is our most common and widespread species.

Applications: Despite its nasty sting, nettle is one of the most delicious and nutritious foods in nature's pantry. Herbalists regard this plant as an excellent nutritive tonic, making it a useful adjunct to virtually any holistic course of therapy or health maintenance. Cooking or thorough drying neutralizes the toxic constituents. Nettle is very high in iron, calcium, potassium, manganese, and vitamins A, C, and D. Young plants are the most palatable, as the plants become tough and fibrous with age. The tea has a folkloric reputation as a menstrual flow regulator and postpartum tonic, a quality attributable to the trace minerals it replaces following childbirth.

Nettle tea is used throughout hay fever season to reduce the severity of chronic, seasonal allergy symptoms. Though scientific validations of nettle's effectiveness in this capacity are questioned by many skeptics, I have met several chronic sufferers of seasonal allergies who claim that it works. They take the herb in the form of tincture, gel cap, or tea (the latter would be my choice) two or three times daily for at least one month preceding the predicted onset of symptoms and continuing through the hay fever season. I have also heard of nettle benefiting dogs that suffer seasonal allergies. Some herbalists theorize that the presence of plant histamines and histamine-like compounds in

Stinging Nettle   *Urtica dioica*

nettle might interfere with the release of histamines produced by the body, thus relieving much of the itching, sneezing, and watery discharge that inevitably results from airborne allergens each spring.

Clinical studies have confirmed the usefulness of nettle root in treating urinary incontinence and swollen prostate, particularly where nighttime sleep is continually interrupted by the urge to urinate.

The mild astringency of nettle makes it an excellent choice for an eyewash and in rinsing the skin and scalp. For more information on the use of nettle in this capacity, see the outline on raspberry.

Alternatives and Adjuncts: Some herbalists use goldenrod as a hay fever preventative. For eyewashes, raspberry, pineapple weed, or chamomile are effective substitutes. For urinary incontinence, also see mullein. Added to virtually any herb formula, nettle is a nutritive bonus.

Propagation and Growth Characteristics: Nettle is a perennial that reproduces mainly from rhizomes and to a lesser degree by seed. Although it may grow in abundance, it is specific and finicky about where it chooses to live. A thick patch of several hundred plants may crowd into a 50-foot strip of stream bank, while the rest of the area is nettle-free for several miles.

With special care you can successfully transplant rhizome cuttings of nettle into the garden. Cultivated nettle is anything but new—North American Indians cultivated this plant as a food, medicine, and cordage crop for thousands of years. To plant nettle, carefully dig and cut the rhizomes in lengths of about 6 inches; if you take too much, you will kill the parent plant. Plant the cuttings 1 or 2 inches deep in soil and habitat that replicates the harvest location as closely as possible. The rhizomes will need to sprawl through the soil freely, and the soil must retain moisture during dry periods. If necessary, amend the soil with copious amounts of peat moss. Keep the transplants evenly moist and they should sprout eventually; don't be surprised if it takes a year or two. Planting nettle in shady, rich soil beneath a dripping garden spigot will make for happy plants. But don't get stung when you hook up the hose!

Gathering Season and General Guidelines: Gather nettle before it begins to bloom in early spring. Once it begins to bloom, the leaves develop cystoliths—gritty particles that can irritate the kidneys. If you gather nettle to eat (it is delicious boiled or sautéed and served with lemon and butter), pick it as soon as it can be positively identified. The young plants typically look rust- or maroon-colored until they are about 6 inches high, and this is the best time to gather them for food. They will turn green as they cook. Be sure to cook them very thoroughly to neutralize their harmful stinging constituents.

To prevent a case of contact dermatitis, wear gloves and a long-sleeved shirt when gathering this herb. Cut the stems at an angle, with a sharp pair of pruning clippers, at least 2 inches above ground level to help prevent microbial infection of the cut stem and rhizomes. For easy handling, insert the cuttings top-down into a paper bag.

If it's the roots you're after, plan a use for the tops as well. Dig the roots conservatively, from the periphery of a healthy patch. Always plant some of the rhizome back into the same spots where each plant was pulled, then mulch the plantings with leaf debris or other nearby materials.

If you get stung, try squeezing some juice from a cut stem of the same plant onto the affected area. Surprisingly, this will often provide immediate relief, as the lecithin contained in the stem juice counteracts the proteins that are partly responsible for the pain. A poultice of plantain or dandelion flower may be helpful. Use the dried herb for teas, soups, and as a food garnish substitute for parsley flakes, or pack it into gel caps for a vitamin and mineral supplement.

Care after Gathering: To dry nettle, use standard procedures (see "Harvesting and Handling Herbs in the Field"). Do not wash the herb before drying it, or it will almost certainly spoil. When the leaves and stems are crispy dry, store them in an airtight container. If the stems feel rubbery, the drying process is incomplete. The dried herb will keep for up to a year, and can be used for tea by steeping in hot water. Use a quantity to suit your taste and be generous—nettle is good food.

ᖋTincture
Fresh or freshly dried plants: 1:2 ratio in at least 50 percent alcohol.

Plant-Animal Interdependence: A thick patch of nettle is an important source of food and protective shelter for reptiles, amphibians, rodents, and ground-dwelling birds. This plant protects delicate riparian habitats by discouraging human encroachment—a feat that deserves applause! Stinging nettle decomposes rapidly and completely, replenishing the soil with substantial quantities of nitrogen, vitamins, and mineral nutrients.

Tread Lightly: The best nettle patches are in moist, humus-rich soil. This means that the delicate, shallow root systems may be vulnerable to the damaging effects of human feet. Also be aware of the likely presence of animals and their dwellings, particularly in dense stands.

> **WARNING! This plant causes an immediate contact dermatitis if handled without gloves and proper clothing. The leaves and stems must be thoroughly dried or cooked before eating!**

Oregon Grape
Berberis (Mahonia) species

Barberry Family
Berberidaceae

Other Names: Mountain grape, mountain holly, western holly, mahonia
Parts Used: Primarily the stolon and attached hairlike roots, and also the aboveground portions of the plant
Actions: Antimicrobial, cholagogue, bitter, and laxative

Oregon Grape is an evergreen perennial that closely resembles holly *(Ilex opaca)*, the Christmas floral. The durable, plasticlike alternate leaves of Oregon grape turn reddish in the fall, have sharp spines at their margins, and glossy upper surfaces that are darker than the undersides.

This plant has juicy, red to purple, ¼-inch berries that grow in narrowly arranged clusters like tiny "grapes." The berries are preceded by bright yellow flowers that form in five whorls, each containing three small flowers. The strong, woody stems form low-growing foliage in most species, with the exception of *B. aquifolium,* which grows as a small, erect shrub. Most species (such as *B. repens*) are stoloniferous ground covers and typically grow in spreading colonies of root-borne offshoots where soils and forest matter are resilient enough for its subterranean sprawl. *B. nervosa* is a horizontal crawler as well, but forms an orderly rosette.

Habitat and Range: Oregon grape inhabits coniferous forests up to timberline from the Pacific Coast to the east slopes of the Rockies north to British Columbia and Alberta and south into the mountains of central California, Colorado, and New Mexico *(B. nervosa* and *B. repens). B. aquifolium* grows mainly in coastal forests and foothills of the Northwest, except where it has been introduced as a landscape shrub.

Applications: This plant contains berberine, a bitter alkaloid that gives the roots their yellow color and a wide variety of medicinal uses. Herbalists use it as a liver stimulant, especially in cases where poor protein metabolism and chronic constipation are seen as symptoms of liver dysfunction.

Oregon grape root can be chewed, or a small quantity of the tincture can be ingested as a before-meal bitter (see dandelion for more on bitters). Dilute a small amount of the tincture or tea in saline solution for use as eyedrops in the antibacterial treatment of conjunctivitis.

Oregon grape is an excellent replacement for goldenseal in applications where an antimicrobial or liver stimulant is indicated. Goldenseal is on the verge of becoming a threatened species because of habitat loss and overharvesting. While we wait for organic farmers to ease market demands for wildcrafted goldenseal, substitute wildcrafted Oregon grape.

Alternatives and Adjuncts: Antimicrobial substitutes might include usnea lichen, bee balm, yerba mansa *(Anemopsis californica),* arrowleaf balsamroot, garlic, or Saint John's wort, depending on the specifics of the infection.

Oregon Grape   *Berberis (Mahonia) aquifolium*

Oregon Grape   *Berberis (Mahonia) nervosa*

Alternative or adjunct cholagogues include dandelion root, chicory root, burdock, red clover, or yellow dock. Oregon grape combines well with virtually any vulnerary (wound healing) herb in cases that warrant an extra-strong antibacterial.

Propagation and Growth Characteristics: Oregon grape is a rhizomatous perennial that grows in large, ground-hugging colonies. One rhizome (a horizontal, subterranean stem) may extend for several yards, sending up a proliferation of offshoots along its length. The plant also reproduces by seeds from the juicy berries. The seeds are small and very hard, and are widely distributed after passing through the digestive tracts of birds, bears, and herbivores.

Stolon cuttings can be established in the herb garden if kept evenly moist throughout their early growth. Many landscape nurseries that specialize in native plants stock Oregon grape as a ground cover. Given its hardy nature (it can survive temperatures as low as minus 50 degrees Fahrenheit), it should be adaptable into virtually any temperate bioregion, but it does require soil that is at least mildly acidic.

Gathering Season and General Guidelines: Oregon grape can be dug anytime, but the roots are most potent in fall; collect them after the plant has gone to seed. Collect the plant by grasping the main stem just above ground level. Pull slowly and steadily upward and the strong stolons will begin to come up out of the ground like a shoelace. When you have exposed a foot or so of root, or if you meet a lot of resistance, clip the root with a sharp pair of pruning shears.

Care after Gathering: Cut the stolons and aboveground parts into small pieces with pruning shears to dry and use in infusions. A better choice though, is to make a tincture.

∾Tincture
Root, fresh and chopped fine: 1:2 ratio in at least 50 percent alcohol.
Dried root: 1:5 ratio in at least 50 percent alcohol.

Plant-Animal Interdependence: This rhizomatous plant is an effective soil aerator and erosion control agent. The berries, which develop late in the season, and sometimes the leaves are browsed upon by deer, elk, moose, bears, and an assortment of rodents.

Tread Lightly: Oregon grape is abundant, widely available, and fairly resistant to human stress. Ever since herbalists discovered that this plant is a viable alternative for goldenseal, one of the most exploited and misused plants in North America, we must be careful that history doesn't repeat itself. Goldenseal once enjoyed a range every bit as large as Oregon grape's current territory, and when we reach for Oregon grape as a goldenseal substitute we need to ask whether using an abundant alternative simply defers the loss to another species.

If we are to save goldenseal, Oregon grape is an important alternative herb. But if you are looking for a substitute for wildcrafted goldenseal (as you should be!), then please consider paying the extra price for organically raised goldenseal instead of reaching for Oregon grape. You will be doing the future of both plants a big healing favor.

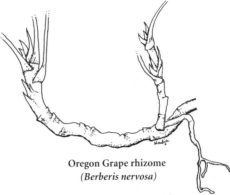

Oregon Grape rhizome
(Berberis nervosa)

Oregon Grape   *Berberis (Mahonia) repens*

Pineapple Weed
Matricaria matricarioides

Sunflower Family
Compositae

Other Names: Wild chamomile, pineapple mayweed
Parts Used: The entire flowering plant
Actions: Carminative, antimicrobial, antispasmodic, anti-inflammatory, vermifuge, and weak sedative

This delightful annual is widely known as "wild chamomile" because it shares a similar appearance and aroma with its close but cultivated relative, chamomile. The primary physical difference between pineapple weed and chamomile is the flower. Chamomile has yellow-centered flowers with white rays (like a tiny daisy), but in pineapple weed flowers, rays are absent. The sweet, pineapple-like aroma of pineapple weed arises at the slightest disturbance of the plant, and once you've experienced it, you'll remember it later when you want to identify the plant again.

Habitat and Range: Pineapple weed ranges throughout the West from central Alaska to California. It must be masochistic, growing in the center of footpaths, highway shoulders, and other high-impact areas where poor soils are always tightly compacted.

Applications: For most purposes, pineapple weed is a wildcrafted substitute for garden varieties of chamomile. Like chamomile, dried pineapple weed flowers make a pleasant tea that is useful for soothing a nervous stomach and for helping to relieve digestive cramping and gas after a torturous food combination.

Pineapple weed poultice, oil, or salve, or a simple water infusion of it, can be applied externally as a soothing anti-inflammatory for insect bites, contact dermatitis, sunburns, and similar ailments. Some herbalists find the same preparations useful as a topical anti-inflammatory for minor varicosity in the lower extremities.

Used in a strong tea, pineapple weed may be useful in expelling round-, whip-, and threadworms from the digestive tracts of humans and domestic animals. If that is true, pineapple weed is a safe and pleasant-tasting alternative to many traditional herbal wormers, such as wormwood (*Artemesia* species) and tansy *(Tanacetum vulgare),* which are potentially toxic to the host as well as the parasites.

Alternatives and Adjuncts: For post–pig-out digestive discomforts, peppermint, catnip, bee balm, or chamomile tea are useful substitutes for pineapple weed. For topical anti-inflammatory options, try plantain, chickweed, cranesbill geranium, aloe vera, or fireweed. For worms, look into the use of pumpkin seeds and garlic. Oregon grape may be useful, too. Pineapple weed makes a good all-purpose first-aid salve when combined with Saint John's wort, plantain, comfrey (or hound's tongue), calendula, or bee balm.

Pineapple Weed   *Matricaria matricarioides*

Propagation and Growth Characteristics: Pineapple weed is a vigorous annual that readily reseeds itself and spreads rapidly. Each plant produces an abundance of diminutive seeds that germinate with a high level of success. The plant requires well-packed soil and tends to do best where it is continually beaten down by people and vehicles. Don't be shy if you decide to transplant this herb. Dig one up during early spring, then stomp it into a hole with the heel of your shoe. Compress the soil around the roots as tightly as possible.

Although moist, shady, low-impact conditions generally yield plants that are attractive, the stronger medicine usually comes from the parched, impact-stunted plants growing in the center of walkways.

Gathering Season and General Guidelines: Gather pineapple weed after the flower heads have turned bright yellow but while the plant is still green. Once you are familiar with the characteristics of this plant, you can judge its potency by the strength of its pineapple-like odor—the stronger, the better. This herb is usually found in high-traffic areas, which means it may be difficult to locate plants that are free of pesticides, gasoline residues, and other contaminants.

Care after Gathering: For teas, the dried flowers of pineapple weed are the most palatable part, but you can use the entire plant. Dry the herb using standard procedures. You can also use the dried herb for making oil infusions or salves.

◯⤳Tincture

Fresh herb, chopped: 1:2 ratio in 40 percent alcohol.

Plant-Animal Interdependence: Birds and many other creatures feed on the sweet flower heads of this herb. At my home, our chickens and the resident grouse and pheasants love it.

Its perseverance in high-traffic areas makes it an important agent in nature's battle against erosion. It provides cover for small animals in areas where none other exists, such as along highway shoulders.

Tread Lightly: Even though a diversity of wildlife relishes this plant, the ecoherbalist must remember that introducing it could be invasive to a natural habitat—don't promote it where it doesn't already exist. Where it grows at the edges of forest trails, pineapple weed is an early-season source of food for hungry birds and small mammals.

Pipsissewa
Chimaphila umbellata

Other Names: Prince's pine, wintergreen
Parts Used: The upper third of the aboveground plant
Actions: Astringent, antibacterial, tonic, diuretic, and nephritic

This small (up to 12 inches tall) evergreen perennial is distinguished by leathery, lance-shaped leaves, which grow directly off woody stems in spokelike whorls. The waxy, almost plastic-looking, pink, ¼- to ¾-inch flowers have five petals and five sepals, terminally borne on stalks extending above the rest of the plant. Pipsissewa reproduces mainly from extensive rhizomes, which form dense colonies of offshoot plants in deep, undisturbed forest compost where the roots can freely creep.

Habitat and Range: Pipsissewa inhabits shaded, coniferous forests where compost is deep and environmental impact (especially in the form of soil

Pipsissewa   *Chimaphila umbellata*

compression) is minimal. In deforested areas or other areas of heavy human use, pipsissewa grows beneath decomposing logs or directly beside a tree—wherever it has peace and penetrable soils.

It ranges from Alaska south to the high mountains of southern California and east across the northern third of North America.

Applications: As a urinary tract astringent, pipsissewa is a much milder and less irritating alternative to uva-ursi, juniper, or alumroot. Herbalists use it to treat urinary tract infections and inflammation of the kidneys, especially where a long-term urinary tract astringent is indicated.

Alternatives and Adjuncts: Pipsissewa is being wastefully exploited. Fortunately, nature has provided us with an abundant herbal ally that is an excellent substitute: pyrola. *Pyrola* species are similar to pipsissewa in pharmacology and identical in usefulness. They are more abundant, more widespread, and faster growing than pipsissewa, and currently there is no commercial interest in this herb. Cranberry juice, too, works as a good, long-term urinary astringent, and so will cranesbill geranium or goldenrod.

Propagation and Growth Characteristics: Pipsissewa is a slow-growing perennial that reproduces mainly from creeping rhizomes and, to a lesser degree, from seed.

Attempts to cultivate this herb have met with failure and frustration. Pipsissewa may require a complex, symbiotic relationship of soil-borne bacteria, fungi, and other natural components to reproduce and thrive. Such balances cannot be artificially replicated, and this makes pipsissewa vulnerable to commercial interests.

Gathering Season and General Guidelines: It should be obvious that I don't approve of harvesting this plant. If you absolutely *must* harvest this plant—say, in the case of an acute bladder infection during a backpacking trip—please look for a useful alternative first. If this fails and your only option is to use pipsissewa, then use a sharp instrument, cut off only the top third of a few plants, and be careful not to disturb or compress the horizontal roots.

Try to gather from the periphery of a large healthy stand, remembering that your entrance into the stand will likely damage the vulnerable soil structures. Avoid walking on mossy or deeply composted soils, and never gather pipsissewa immediately after rainy periods. Please—take only what meets your immediate needs.

Care after Gathering: Use the fresh herb to make a strong decoction in emergency situations, a handful of chopped herb to 1 quart of water. Pipsissewa does not take to glycerine.

Plant-Animal Interdependence: Like many rhizomatous, deep-forest plants, pipsissewa is important as a soil aerator. The long, horizontally crawling roots break up the dense, carbonaceous forest debris so air, water, and organisms can get in—the essential elements needed to nourish and sustain soil and plant life.

Tread Lightly: Although the present and potential threat to pipsissewa is ominous, most herbalists are unaware of how close the herb is to disappearing. The predicament is mainly because pipsissewa is an ingredient in certain types of soft drinks. A reliable source informed me that a large soft drink company is hauling pipsissewa out of the forests of Oregon and Washington by the container load, with little or no regard for environmental protection or reproductive sustainability. In the Cascade Mountains, many areas that once supported vast stands of pipsissewa are now relatively void of the plants. Adding to the frustration that elevates my blood pressure whenever I write or speak of this issue is the fact that pipsissewa doesn't even taste good. For the soft drink companies to obtain the flavor they want from this plant, they have to extract a specific, isolated compound from its diverse chemistry, which means they must use an enormous number of plants to extract a small amount of usable product. This mass harvesting, coupled with the delicate, slow-growing nature of the plant and its specialized habitat, leaves pipsissewa and its environment in a precarious position.

# Plantain
Plantago major

Plantain Family
Plantaginaceae

Other Names: Common plantain, English plantain, hoary plantain, psyllium
Parts Used: Any or all parts of the entire plant
Actions: Emollient, demulcent, astringent, nutritive, and laxative

Common plantain is an annual or perennial herb characterized by a low-growing rosette of broad leaves and a drab but distinctive flower cluster. The succulent but sturdy leaves are on proportionately long leaf stems (petioles) and have distinct parallel veins that contain strong fibers. The flowers are tiny and inconspicuous, borne in tightly arranged sausage-shaped spikes atop leafless stalks that reach well above the rest of the plant. Leaf configurations vary between species, from egg-shaped *(P. major)* to narrowly linear *(P. patagonica, P. psyllium,* and *P. elongata)*. Most plantain species share a fairly similar appearance, notably in the terminate flowers atop leafless stalks.

Habitat and Range: Common plantain prefers high-impact areas and typically grows in the center of dirt roads and walkways, even in cracks in highways. Ten or more *Plantago* species inhabit western North America, with common plantain *(P. major)* by far the most widespread and abundant.

Applications: Plantain is effective as an emollient when used as a fresh poultice on insect bites, minor burns, and other skin irritations. Herbalists use the juice of the plant to soothe intestinal irritations, hemorrhoids, and even stomach ulcers. The seeds used fresh in tea act as an effective, lubricating laxative. The seed husks of *P. psyllium* are used for the same purpose in many over-the-counter dietary fiber supplements and laxative preparations. Plantain tincture or tea is sometimes used to soothe a rough, irritated throat, and may be useful in easing the passing of gravel from the urinary tract.

Much of the medicinal usefulness of plantain derives from the plant's mucilage content. Mucilage is a compound with the consistency of sticky motor oil, making it helpful when a soothing, lubricating, protective barrier is needed to bring symptomatic relief to physical complaints both internal and external.

Alternatives and Adjuncts: Plantain may be a good substitute for slippery elm bark *(Ulmus fulva)*, a tree with mucilaginous and astringent properties that is disappearing because of irresponsible wildcrafting practices, commercial logging, and Dutch elm disease.

Marshmallow or gravel root are better choices than plantain for respiratory, digestive, or urinary irritations or infections. Fireweed, aloe vera, cranesbill geranium, or chickweed are alternatives to consider for many topical and internal applications. In topical first-aid preparations, plantain combines with Oregon grape, usnea lichen, bee balm, or Saint John's wort.

Common Plantain   *Plantago major*

Plantain *(Plantago lanceolata)*

Plantain flowers
(Plantago lanceolata)

Propagation and Growth Characteristics: Plantain grows as a perennial or annual, depending on species, elevation, and climate. It reproduces mainly by seed and, to a lesser extent, from perennial division. The introduction of this herb into the garden should be pursued only by genuine weed lovers, as it will find its way out of the flowerbeds and into pathways, driveways, and perhaps the neighbor's yard. Plantain is not picky about soil quality, is highly drought tolerant, and prefers sunshine to shade.

Gathering Season and General Guidelines: Gather plantain leaves and roots anytime, so long as they are green and healthy looking. Many people regard plantain as a weed, so gather from an area that has not been treated with herbicide. If you are not yet familiar with this plant and plan to gather it from atypical, low-impact areas, you should probably wait until it is flowering. Many members of the Lily family can be mistaken for plantain during early growth, and some of those mistakes could prove toxic if ingested. Once plantain blooms, its identity is almost unmistakable.

Gather plantain by grasping the plant at its base and pulling. This sounds simple enough, but in dry soils, the aerial plant will break off at ground level. If you want the root, use a ditch spade or hand trowel.

Care after Gathering: This herb is best used fresh. For bites, stinging nettle attacks, and first-degree burns, a first-aid poultice can be made quickly by chewing some leaves and applying the green goo directly to the affected area. The plant is entirely edible and quite tasty. At home, use a blender or juicer to make topical poultices or purees for internal uses.

Plant-Animal Interdependence: Plantain is a standby staple for herbivores. It is seldom a primary forage but is usually available. Its capacity to survive drought and continual trampling make it an effective soil structure engineer. Plantain's roots, though not always as fibrous and tough as the leaves, effectively penetrate hardpan soils and prevent erosion in such areas as road margins.

Tread Lightly: This plant is an indicator species for surrounding conditions. If a stand of plantain appears to have been foraged, be suspicious that other sources of forage may have been depleted. Don't harvest it if the animals need it more than you do.

Poplar

Populus species

Willow Family

Salicaceae

Other Names: Black cottonwood, quaking aspen, white cottonwood
Parts Used: The leaf buds and the inner bark (cambium)
Actions: Anti-inflammatory, vulnerary, antimicrobial, and expectorant

Poplars are deciduous trees or large shrubs characterized by catkin-type uni-sexual flowers; sharply pointed, broadly lanceolate to nearly ovate, long-peti-oled leaves; and the consistently moist habitats where they live. The drooping male and female catkins are borne on separate plants. Black cottonwood *(P. trichocarpa)* and quaking aspen *(P. tremuloides)* are among the most widely distributed and readily recognized species in North America, and are excel-lent examples on which to base your familiarity with the rest of the *Populus* clan. Black cottonwood is a large tree (up to 120 feet) that stands out in its riparian habitat. The leaves are broadly oval, with distinctively pointed tips; the undersides are almost white, with rust-colored veins. The bark of the young tree is white and smooth, later becoming gray brown, thick, and deeply fur-rowed as the tree matures. Quaking aspen is even more distinctive, with nearly circular, pointed leaves that tremble ("quake") in the slightest breeze, creating a unique rustle that, from a distance, sounds like running water. The greenish white bark stands out against surrounding foliage.

In fall, when poplar leaves turn bright golden yellow, these unique splen-dors meld to create a Thanksgiving feast for the senses.

Habitat and Range: Poplars inhabit moist woodlands and riparian areas across most of western North America. Black cottonwood typically grows along riverbanks and in wet floodplains up to about 5,000 feet. Quaking aspen is an upland tree, inhabiting forest clearings up to timberline. All poplars require

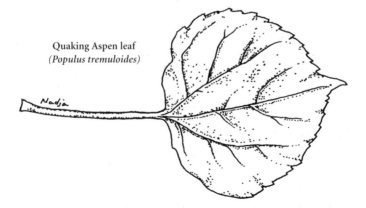

Quaking Aspen leaf
(Populus tremuloides)

Black Cottonwood   *Populus trichocarpa*
Inset: Black Cottonwood bud covered with balm of Gilead   *Populus trichocarpa*
Black Cottonwood   *Populus trichocarpa*

an ample and consistent water supply. If surface water is not visible near one of these trees, you can safely assume the roots are being fed by a subterranean source.

Applications: The primary action of poplar is anti-inflammatory, used both topically and internally. In early spring, the leaf buds are covered with a sticky, red resin, known by herbalists as "balm of Gilead," which has a broad array of medicinal activities. My favorite uses for balm of Gilead are as an ingredient in lip salve and for relieving pain and inflammation in the mouth or throat, such as tonsillitis or the post-trauma of having had a tooth pulled. For either of those, I use it as a gargle, made from a tincture diluted in salt water.

The inner bark of this tree is a reliable source of populin and salicin, compounds that act like crude forms of aspirin, except that they will not help thin the blood as aspirin does.

Alternatives and Adjuncts: For use in healing salves, poplar combines well with Saint John's wort, plantain, Oregon grape, calendula, or pineapple weed (or the related chamomile). As an analgesic or anti-inflammatory, willow is a useful substitute. For the throat and mouth, bee balm and cow parsnip seed are options worth considering.

Propagation and Growth Characteristics: Poplars are deciduous trees that require a constant source of moisture. These trees are a beautiful addition to a yard or pasture and are easy to establish from nursery stock. All you need is an adequate water supply. Poplars reproduce from root runners or by seed, and many species can be propagated simply by placing cuttings of the young branches into a glass of water until they take root.

Gathering Season and General Guidelines: The leaf buds yield the most potent medicine and can be gathered with minimal ill effects. Collect the young, sticky leaf buds before they are halfway open, in early to midspring: if the red balm of Gilead is gone, you are too late. Gather conservatively from branches within easy reach, taking care *not* to pick the terminal end buds, which are essential for limb growth. Carry some rubbing alcohol with you to clean your fingers, otherwise you will soon find that you cannot release the sticky buds from your grasp!

Gather poplar bark during spring, when it separates easily from small limbs. To remove it, make a small incision with a sharp knife and peel a narrow strip of bark from a few limbs within easy reach. If a gooey glob of pine pitch is available nearby, use it to cover the wound that you have caused. A commercial brand of pruning compound will also do. Another, perhaps better, option is to prune a few of the small, pencil-sized outer limbs from the tree. You can cut them into small pieces and tincture them, avoiding permanent scarring of the tree.

Care after Gathering: If gathered early enough in the spring, the moist inner bark can be peeled away from the outer bark and dried. This has a higher potency-to-bulk ratio and is better tasting in teas than the outer bark. If you can't separate the bark, dry it and use it as is. The dried bark will store in an airtight container for a year or more.

The leaf buds are best used fresh. The resins you need to extract are not water soluble, so the buds must be tinctured before the balm of Gilead can be added to oils, ointments, or salves.

∿Tincture

Fresh leaf buds or fresh bark or small limbs: 1:2 ratio in at least 50 percent alcohol. Poplar does not take to glycerine.

Plant-Animal Interdependence: Poplars are food, shelter, and nesting habitat for many organisms, from pollinating insects and ants to ospreys, eagles, and large mammals. Whitetail deer and moose browse the lower leaves and young branches and, during hard winters, forage the rich, inner bark. Cottonwoods and aspens are favorites of beavers.

Tread Lightly: Poplars are a major component of riparian habitats, where they are critically important to birds, mammals, amphibians, and other organisms. The streamside habitats where poplars grow, though, are also desirable building sites for people, and an ever-increasing number of cottonwood and aspen stands are falling to the chain saws of developers, home owners, and firewood gatherers. Poplars regenerate quickly but the diverse and tightly interdependent riparian habitats may never reestablish once they are gone.

Next time you enter a poplar thicket, take time to look beyond what they may represent for the human body and consider what they represent for the human condition. When the poplar thickets are gone, so, too, will be precious opportunities to understand the ways of our planet. When you are in these special places, close your eyes, breathe deeply, taste the air as it fills you with the sweet aroma of life, and let your imagination blend with the natural beauty around you. Then, you will know what holistic healing with plants is all about.

WARNING! Many people are highly allergic to this plant.

Pyrola
Pyrola species

Heath Family
Ericaceae

> **Other Names:** Wintergreen, pink pyrola, one-flowered pyrola, wood nymph
> **Parts Used:** The leaves
> **Actions:** Astringent, antibacterial, diuretic, and nephritic

Pyrola earns the common name "wintergreen" because of its cold-hardy nature. These perennial evergreens stay in choice condition all winter, even beneath a blanket of snow in subzero temperatures. *P. asarifolia* is one of the largest (up to 12 inches tall) and most widespread pyrolas. Like many other broad-leafed species, it has nearly circular, glossy, dark green, petiolate leaves arranged in flattened, basal rosettes. Large groups of pyrola typically carpet the forest floor in densely spaced, ground-hugging colonies, the result of the plants' extensive rhizomes sending up shoots from the many nodes along their lengths. The flowers each have five petals and form on a leafless stalk rising well above the rest of the plant. Although the flowers also distinguish the plant, they tend to bloom only a few at a time and may be inconspicuous among other vegetation.

Habitat and Range: Pyrolas require a mysterious balance of fungi and other soil constituents. They are most abundant in shaded, moist forests where dead wood and deep compost are plentiful, especially where humans have not encroached. Several species are widely distributed across the Northern Hemisphere.

Applications: Pyrola is an excellent astringent and disinfectant, especially useful for urinary tract infections. It is pharmacologically similar to its heavily overharvested relative, pipsissewa, and can be applied in exactly the same manner for inflammations of the urinary tract.

Alternatives and Adjuncts: Where a weak astringent is indicated (eyewashes, for example), raspberry leaf, bee balm, cranesbill geranium, or nettle are better alternatives. Where a strong astringent is needed in acute situations (such as poison ivy dermatitis or treatment of hemorrhoid flare-up), uva-ursi, alumroot, or juniper are better choices.

Propagation and Growth Characteristics: Pyrola is a perennial that sends up offshoots from nodes along creeping stems and roots. It also reproduces by seed. This plant does not transplant well and is nearly impossible to start from seed. The natural habitat that pyrola enjoys is one of incomprehensibly delicate symbiotic checks and balances that cannot be mixed up in a garden bucket.

Gathering Season and General Guidelines: Pyrola is useful anytime, but you should gather it after its reproductive cycle is complete, in early fall. Find

Pyrola   *Pyrola asarifolia*

a large patch with easy access and cut a leaf or two from several healthy plants. Avoid *pulling* the leaves—the roots are weak and easily damaged. Also, the spongy, compost-rich soil where pyrola grows is easily compressed by human feet. Wear soft shoes and plan your approach carefully.

Care after Gathering: The fresh leaves can be made into tincture or they can be used in a decoction. Air dry the leaves for future use.

∾Tincture

Fresh leaves: 1:2 ratio in 50 percent alcohol.

Plant-Animal Interdependence: A close look at pyrola and its environment reveals a wealth of information about how ecosystems operate. The longer and harder we study this plant, the more we learn about the complex interdependency of everything surrounding it. We see that fungi and other microorganisms break down dead plants into nutrient-rich soil. We surmise that pyrola seed is dispersed by foraging animals or perhaps by water runoff. After establishing a foothold, pyrola then begins working as a steward of the soil—its extensive rhizomes prevent erosion, hold nutrients, and house insects and microorganisms. We discern that pyrola cannot contribute these services without the continued assistance of other organisms—the insects that pollinate it, the animals that feed on it, the beaver that trims the willows and lets in just the right amount of sunlight through the forest canopy. It's an interdependent continuum.

Tread Lightly: Although this herb is plentiful and widespread, it grows in an environment that is vulnerable to human interference. Only harvest this plant for use as a gentle urinary tract astringent. For other situations where an astringent is indicated (particularly a stronger one), use one of its many alternatives.

Raspberry
Rubus species

Rose Family
Rosaceae

> **Other Names:** Red raspberry, blackcap raspberry, bramble-berry, thimbleberry, salmon berry, black berry, black raspberry
>
> **Parts Used:** Dried leaves and the fruits
>
> **Actions:** Astringent, uterotonic, nutritive, diuretic, laxative, and mild sedative

Wild raspberries are generally categorized by color, red or black. Leaf and stem characteristics vary between species of this widespread genus of shrubs. Most have pinnately divided leaves and five-petaled flowers ranging from white to crimson. They typically grow as tangled masses of thorny, trailing biennial stems that yield a tasty reward to those willing to brave the thorns to reach the choicest berries. Although flavor quality and size varies from species to species, the fruits look essentially the same as the cultivated, market varieties. The fruits generally are borne from axillary or side-branch flowers. Thimbleberry *(R. parvifloris)* is the most unusual-looking member of the *Rubus* clan,

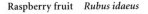

Raspberry fruit *Rubus idaeus*

with large (2- to 6-inch), lobed, maplelike leaves, no thorns, and terminal, main-stem clusters of two to three flowers or fruits.

Habitat and Range: As you travel north through the Rocky Mountains or the coastal states of the West, the diversity and number of roadside "bramble patches" progressively increases. Many *Rubus* species are regarded as invasive weeds. They are a common and often abundant inhabitant of pastures, roadside ditches, and riparian habitats. This genus often cross-pollinates, sometimes making identification of species difficult. Many species have escaped cultivation, and it's anyone's guess where a berry plant might pop up next. Red raspberry *(R. idaeus)* is one of the most common indigenous species across North America.

Applications: Raspberry leaf is a very safe and gentle food medicine. Aside from its nutritional value (especially vitamin C), raspberry leaf tea has mild astringent qualities that make it useful for many remedial and tonic therapies. Taken internally, the tea relieves minor digestive tract inflammation and mild cases of diarrhea. The leaf tea is perhaps the most widely used "female tonic." Its actions are said to improve the tone and elasticity of smooth muscle tissues in and around the uterus.

In my experience, raspberry leaf tea is especially useful as a soothing eyewash, for symptomatic relief of conjunctivitis. I use raspberry as a substitute for eyebright *(Euphrasia arctium),* a plant that is currently at risk of disappearing because of overharvesting. The cooled, full-strength tea also makes a soothing scalp or skin rinse; the astringency will help relieve minor itchiness.

Alternatives and Adjuncts: Strawberry leaves (*Fragaria* species) are a good substitute for raspberry leaf. As a "female tonic," nettle leaf tea will substitute, with a richer array of nutrients. Nettle, too, can be used for eye, scalp, and skin rinses. Raspberry leaf tea is delicious, especially when combined with rose hips, hibiscus flower, huckleberry leaf, a squirt of lemon, and some honey.

Propagation and Growth Characteristics: Like cultivated varieties, wild raspberries produce long, tangled biennial canes that produce fruit on second-year growth. The fruits of wild raspberry are smaller than domesticated species.

Although wild raspberry typically grows in dense, thorny, impenetrable thickets in disturbed roadside areas and vacant lots, it is difficult to establish from cuttings or transplants. Cultivars are a better choice for your garden.

Raspberries require acid soils, ample moisture, and lots of potassium (potash). Although they flourish just about anywhere, plants in full sun will yield more fruit than those in shade.

Gathering Season and General Guidelines: For the strongest medicine and best-tasting tea, gather leaves just as the flower buds are forming in early spring. If you miss this harvest date, the older leaves can be used if they are in good condition. Wear gloves and a long-sleeved shirt to protect yourself from the prickly stems, and gather from the periphery of several different patches. Avoid entering any berry patch, which can result in blood loss (sharp thorns!) and may disrupt creatures that use the bramble-protected thicket for home and hiding. Gather fruits when they are ripe, in the same manner as the leaves. In areas where people treat raspberry species as troublesome weeds, beware of herbicides.

Care after Gathering: Use raspberry leaves only after they are completely dried or when they are fresh off the plant, not when they are in the process of drying; the leaves may temporarily develop mildly toxic substances as they wilt. The level of toxicity is not life-threatening, but the wilted leaves could induce nausea.

I use dried leaves for tea, as they are more soluble and better tasting than fresh. After the leaves are dry and brittle, store them in an airtight glass jar away from direct sunlight. The dried herb will keep for a year or more.

To make an eyewash, infuse 1 teaspoon of dried raspberry leaf in 8 ounces of hot, distilled water. Allow the tea to cool, strain it through a coffee filter, then combine it, a little at a time, with sterile saline until the saline is obviously tinted but still transparent.

Plant-Animal Interdependence: As anyone could guess, raspberry fruit is relished by everything from mice to bears. Birds are particularly fond of the berries; migratory flocks return to the same bramble patches year after year.

A dense, thorny entanglement of raspberries provides protective cover and habitat for birds and small animals. In the streamside patches near my cabin, you are almost guaranteed to find grouse nests there in the spring and early summer months.

Tread Lightly: Raspberry provides critical habitat and food for wildlife, and the removal of large stands would profoundly affect the biodiversity of its environment. Because of its spreading habits, be careful not to introduce this plant where it is unwelcome.

Red Clover
Trifolium pratense

Pea Family
Leguminosae

Other Names: Crimson clover
Parts Used: All aboveground parts of flowering plant
Actions: Alterative, nutritive, and antitumor

Red clover was introduced from Europe as a livestock feed, then escaped cultivation and made itself at home across the continent. It has the typical predominantly tree-lobed leaves of the *Trifolium* genus, which has dozens of species and cultivars distributed throughout the West. The differences among these many species may be slight—positive identification of a specific *Trifolium* can be a challenge. Begin identification of red clover by noting the red, globe-shaped flowers and the softly hairy stems. The flower stems (pedicels) are shorter than each of the leaf stems (petioles). Red clover has a taproot, though many clovers have rhizomes.

Habitat and Range: Red clover is widespread in cultivated fields, road margins, gardens, and any other disturbed area where the plant has been introduced through agriculture.

Applications: Herbalists know red clover is a traditional "blood purifier," an herb that helps free the blood of toxins and systemic waste while providing an assortment of nutrients critical to healthy blood. The term *blood purifier* is erroneous, because red clover doesn't "clean" the blood so much as nourish it to a healthy state by providing essential nutrients. It also gently supports the liver in helping it produce red blood cells and filter waste. Red clover is a "medicinal food."

Red clover is especially rich in iron and vitamin K, as well as coumarin, a compound known to act as a blood-thinning agent. Scientific studies suggest that red clover has tumor-inhibiting properties as well, which substantiates its reputation among herbalists as a long-term supplemental therapy for cancer patients.

Alternatives and Adjuncts: For a general nutritive, nettle is a better choice. Herbalists often combine red clover with cholagogues and alternatives such as dandelion, burdock, milk thistle, or yellow dock for use in liver detoxification formulas.

Propagation and Growth Characteristics: If you have rich soil and an ample water supply, you'll be able to propagate red clover with great success. It is an excellent green manure plant that will contribute nitrogen and trace minerals to the soil when tilled in after you've harvested the flowers in midsummer. Watch out, though—this perennial can become a nuisance in areas of the garden where you have better things to do. Red clover seed is available through nurseries and catalogs that specialize in cover crops.

Gathering Season and General Guidelines: Harvest red clover when it is in full bloom, usually between early June and mid-August. The blossoms (without leaves and stems) are the most desirable portion of the plant, containing the heaviest concentration of medicinal constituents. The entire plant is useful, though, and unless you are harvesting it for a commercial manufacturer who insists on "flowers only," there's little reason to engage in the tedious task of individually picking hundreds of blossoms to make a tincture. With a pair of grass clippers, cut the top 2 to 3 inches of the flowering plants—leaves, stems, and all. If it's a delicious tea you're after, then go ahead and hand pick the flowers, which taste better than the stems and leaves.

Avoid gathering in areas adjacent to cultivated fields, road margins, or other places where herbicides and other toxins might be present.

Care after Gathering: Use standard drying procedures for herbs. The flowers will feel papery when they are completely dry, and the leaves will crumble

Red Clover *Trifolium pratense*

between your fingers. The dried herb will keep for a year if properly stored (in plastic bags or glass jars) out of direct sunlight.

∿Tincture

Fresh herb, chopped: 1:2 ratio in at least 40 percent alcohol. *Dried herb:* 1:5 ratio in 40 percent alcohol.

Plant-Animal Interdependence: Red clover is an effective pollinator attractor, drawing insects down from the high canopy to ground level, where they subsequently visit other ground-hugging plants. Although red clover is an alien species in all North American biosystems, it contributes a rich array of nutrients to the soil. Virtually any herbivore relishes the nutritious flowers and foliage. In ecosystems where natural balances have been disrupted, red clover typically shows up as a supplemental food source for animals.

Tread Lightly: Because red clover is not a plant native to North America, be careful not to introduce it into habitats where it does not already exist. When harvesting it from the wild, leave reproduction to nature's discretion; an artificially induced proliferation of red clover might compete with native species.

Rose
Rosa species

Rose Family
Rosaceae

Other Names: Wild rose
Parts Used: The hips, petals, leaves, and roots
Actions: Astringent, nutritive, diuretic, and vulnerary

Wild roses look and smell like their domesticated counterparts, but their flowers and leaves are smaller. Characterized by pink, five-petaled flowers, thorny stems, and bright red to purplish fruits (rose hips), this plant is a safe, easy-to-identify, sensually delightful introduction to nature's pantry and apothecary.

Habitat and Range: Wild rose grows in consistently moist soils, typically in dense thickets at road margins, in irrigation ditches, and especially as the defensive edge of riparian habitats, up to about 6,000 feet. Several species are native across North America. Wood's rose *(R. woodsii)* is one of the most widespread species in the West.

Applications: Most people today know about the lovely flavor and vitamin-C richness of rose hips, but over the years we've lost knowledge about the medicinal usefulness of this huge plant genus. Wild roses are widespread and abundant, and we can use them instead of dozens of other medicinal plants that are far more vulnerable to human interference. Wild roses (and domesticated varieties, too) can be made into an effective astringent medicine with several levels of potency. The flowers are mildly astringent, and the petals can be plucked off and applied directly to an insect bite or minor injury as a membranous, soothing, and protective Band-Aid. The leaves—the next step up in astringency—can be made into a tea for use against minor bouts of diarrhea or as an astringent skin rinse for minor cases of contact dermatitis or an itchy scalp. If you need an even stronger astringent—to stop bleeding in the upper gastrointestinal tract, to relieve inflamed hemorrhoids, to remedy a urinary tract infection, or as an anti-inflammatory mouthwash and gargle—make a decoction with the cane bark or small twigs. If this isn't strong enough, you can get even more astringency from the high-powered root bark. Whenever an astringent is indicated, one wild rose bush can serve the same purposes as cranesbill geranium, pyrola, or uva-ursi. Add a mucilaginous herb, such as marshmallow *(Althea officinalis),* and you have a digestive tract alternative to slippery elm *(Ulmus fulva),* a deciduous tree of eastern North America that is disappearing under the combined pressures of herb industry demand and Dutch elm disease.

Alternatives and Adjuncts: Uva-ursi, pyrola, cranesbill geranium, or virtually any of the astringent herbs are effective substitutes for wild rose, depending on what astringency the symptoms call for.

Wood's Rose   *Rosa woodsii*

Wood's Rose hips   *Rosa woodsii*

Propagation and Growth Characteristics: Wild roses are perennial shrubs that reproduce easily from roots and seeds. They hybridize with one another readily and will not hesitate to seduce open-pollinating domestic roses. You can easily start wild roses in the garden, either from stratified seed or by transplanting rootstock (available in many nurseries). Give the young plants full sun and plenty of water until they are well established, and you have them forever. Remember: these thorny wild species may take a liking to other Rosaceaes in your garden and hybridize with them.

Gathering Season and General Guidelines: Gather rose hips after the petals have fallen and the hips have turned bright red. The tastiest and strongest hips are slightly shriveled, and have been through at least one fall frost. The hips stay in good condition on the bush well into winter but develop mold soon after they shrivel, so don't wait too long. The fresh hips make excellent jellies and wine and are quite tasty eaten fresh, despite the hard seeds.

The flowers can be used anytime. The bark is best if collected in early spring, when the energies and chemistries of the plant are being directed out of the root and into the budding canes. To harvest the bark, use a sharp pair of pruning shears to harvest a couple of the young canes, and then whittle the bark off with a sharp knife (wear thick gloves, and beware of the thorns!). If you don't need bark-strength medicine but the flowers are not a useful option, collect the small, end twigs and leaves. Clip them off conservatively and disperse your harvest among several plants. If you need the potent root bark, carefully dig around the base of the plant until you find a small side-root; if you work slowly and carefully, you can remove it without causing permanent harm to the plant. Another option is to peel a small strip from one side of the main root, but keep it small or you will kill the plant.

Avoid gathering from roadsides and train easements as plants there probably contain toxic residues.

Care after Gathering: Dry the hips following the standard procedures in "Harvesting and Handling Herbs in the Field." Be certain that the hips are completely dry before storing them in an airtight container. Properly stored, away from light, they will keep well for about a year.

To make an infusion (tea), crush some of the dried hips and steep them in very hot water. Adjust the quantity of hips to taste. The leaves can be used fresh in teas or dried and stored by the same method.

The bark can be used fresh, but may cause a reverse, laxative effect if used in excess. I prefer to dry it. Again, spread the bark shavings on a clean piece of paper and stir them once in a while to prevent mold growth. The bark can be used to make a weak tea or can be decocted into a stronger medicine. For details on making a decoction, see "The Basics of Making Herbal Preparations."

Plant-Animal Interdependence: Rose hips are relished by birds, deer, coyotes, bears, and anyone else who discovers their chewy sweetness. Even my German shepherd browses them when given the opportunity. Thick, entangled, thorny stands of wild roses provide protective cover and habitat for a wide variety of insects and animals. Bees and other pollinators love the bright, fragrant flowers.

Tread Lightly: Always check for possible wildlife inhabitants when you gather from a stand of wild roses, and leave plenty of fruit for the critters who depend on it. This valuable food and medicine plant may be condemned by cattle ranchers, farmers, and garden enthusiasts because it is so free-spirited and thorny.

Sage
Salvia species

Mint Family
Labiatae

Other Names: Black sage, white sage, purple sage, sagebrush, wild sage

Parts Used: All aboveground parts

Actions: Antiseptic, astringent, hemostatic, anti-hidrotic, alterative, and tonic

Wild sages are highly aromatic, with a fragrance much like the culinary variety *(S. officinalis)* but stronger. The aroma is not unique to *Salvia* species, however, and may be misleading for identification unless you study other characteristics. For instance, big sagebrush, wormwood, and dozens of other *Artemesia* species have a fragrance similar to sage, but the *Artemesia*s belong to the Sunflower family. The main point to remember for differentiating an *Artemesia* from a *Salvia* is that all members of the *Salvia* clan have opposite

White Sage   *Salvia apiana*

Purple Sage   *Salvia leucophylla*
Inset: Purple Desert Sage   *Salvia dorrii*

leaves and, in most species, four-sided stems. In the *Artemesia* genus, the plants have alternate leaves.

Many species of wild sage share the unique "pebbled" leaf texture characteristic of their cultivated counterpart. The leaves and stems of many species are fuzzy, giving the plants a blue gray appearance. The flowers range from white to deep purple, depending on species, and form in whorled clusters where the leaves join the upper stem or in whorled, terminal spikes, again depending on species.

Habitat and Range: Several sages inhabit the canyons and foothills of the western United States; the greatest variety of native species is in the Southwest. In the coastal canyons of southern California, white sage *(S. apiana)*, black sage *(S. mellifera)*, Munz's sage *(S. munzii)*, and purple sage *(S. leucophylla)* grow as the predominant flora in dense stands of 2- to 6-foot-high shrubs that may cover the landscape for miles. As you travel north or east, the distribution of *Salvia* species becomes more scattered and less varied, with plants generally small and less predominant in their habitat. In the deserts of southern Idaho, Utah, the northern reaches of Nevada, and the Columbia Gorge of Oregon and Washington, purple desert sage *(S. dorrii)* lends colorful contrast to its habitat. In the Midwest and eastern United States, lyre-leafed sage *(S. lyrata)* grows in dry, sandy soils. Check your local plant key or herbarium for the sage nearest you!

Applications: While specific uses for sages are as varied as the *Salvia* genus, most are similarly useful in their general range of application. All sages have antiseptic qualities. I find the fresh or dried leaf poultice or tea especially useful for infections or ulceration of the mouth. The leaf tea also can be useful as an antimicrobial and astringent skin and scalp rinse; use it when excessive, perhaps nocturnal, scratching has compounded the itch of insect bites by producing infected, open sores and scabs (if you wonder what I'm talking about, spend a summer night in a Montana willow thicket without a tent!). An itchy, flea-bitten dog might enjoy a sage rinse. Make an instant tea by placing a teaspoon or more of the leaf tincture into a cup of warm or cold water.

White sage *(S. apiana)* is a traditional ceremonial herb of Southwest Indian cultures, who use sage smudges to cleanse the body and spirit of physical and nonphysical impurities. Today, white sage smudges have become a popular item in New Age bookstores and herb shops, and thousands of people are using the herb as a sort of New Age air freshener. I find this herb particularly powerful for enhancing my mental clarity during quiet meditation; to me, it is a sacred herb. Out of respect for the plant and its habitat, however, I seldom use it. This herb deserves to be saved for special ceremonies, as it has been for thousands of years.

Alternatives and Adjuncts: Rosemary *(Rosmarinus officinalis)* works well as an antiseptic and astringent substitute for sage. Bee balm *(Monarda* species) is an excellent alternative for mouth infections and sore throats.

Propagation and Growth Characteristics: Some sages grow as annuals, others as biennials or perennials. Most sages are drought tolerant and winter hardy, and prefer soils low in organic matter. Besides the culinary varieties, dozens of native species are available through nurseries, particularly ones that specialize in native landscape plants.

Gathering Season and General Guidelines: Sage can be gathered anytime, but the strongest herb comes from blooming plants. The young, stem-tip shoots are the best-tasting part of the plant and are the first choice for teas. Selectively and conservatively pluck the tips, the leaves, or cut the flowering stems.

Some sages are frowned on by livestock owners and herbicide enthusiasts. Lanceleaf sage *(S. reflexa)* is said to contain high levels of nitrates and reputedly is toxic to livestock. Mediterranean sage *(S. aethiopis)* forms tumbleweeds, and some people regard it as troublesome enough to warrant chemical warfare. Check for herbicides before you gather.

Care after Gathering: Use the leaves, stems, and flowers fresh or dried in infusions and poultices.

∾Tincture

Dried herb: 1:5 ratio in 50 percent alcohol.

Plant-Animal Interdependence: Sage is forage for many animals, but it is more important as a source of cover and habitat. This is especially true where sage grows in abundance, as in the coastal foothills of southern California. Beneath those dense herbal canopies lives a diverse biocommunity of rodents, reptiles, insects, and ground-dwelling birds. Many creatures spend their entire life inside a single stand of the shrubs—eating, sleeping, reproducing, and dying within the protective and aromatic cover of *Salvia* species.

Sage is also an effective pollinator attractor. It is popular with bees and beekeepers alike—nothing compares with the tangy sweetness of sage honey.

Tread Lightly: Most sages are in no danger of disappearing, but many of the sage-covered canyons I played in as a child are now covered with asphalt and stucco, and many of the animals that depend on these plants are vanishing. Natural abundance is a gift to be appreciated, not taken for granted. The next time you gaze out on an expanse of sage, please consider what the future may hold.

Saint John's Wort
Hypericum species

Saint John's Wort Family
Hypericaceae

Other Names: Klammath weed, goat weed, Fourth of July flower
Parts Used: The top 12 inches of the flowering and/or postflowering plant
Actions: Vulnerary, nervine, antidepressive, antiviral, bacteriostatic, anti-inflammatory, and astringent

Saint John's wort is a sturdy perennial weed distinguished by its yellow, five-petaled flowers, each with numerous stamens, and its small (up to ¾-inch), narrowly lance-shaped to elliptical, opposite leaves. The flowers and leaves are covered with tiny, purplish black glands that contain hypericin, a medicinally active compound visible as a red stain on the skin after rubbing the foliage between your fingertips. The size of the plants varies between species, but all look similar. The largest and most widespread species, *H. perforatum*, may reach 32 inches tall, though bog Saint John's wort *(H. anagalloides)*, one of the smallest species, grows in mats and seldom exceeds 4 inches. Sometimes referred to as Fourth of July flower, Saint John's wort usually blooms from early July through August.

Habitat and Range: Its habitat varies according to species, but generally the larger species prefer dry to moist, open hillsides up to 6,000 feet. Higher elevations and wetter habitats support the smaller species. In the Pacific Northwest, *H. perforatum* and *H. formosum* are common and may grow in profusion on open rangelands at foothill elevations, where they are regarded as noxious weeds because of their alleged toxicity to livestock. Saint John's Wort is a European import that today is abundant in many regions of the Pacific and Rocky Mountain states and sporadically ranges across the rest of North America.

Applications: Saint John's wort is well known for its remarkable ability to stop infection, reduce pain, and speed the healing of burns and wounds. Many herbalists use it as a specific remedy for various strains of herpes virus, sciatica, and neuralgia.

Using Saint John's wort to treat chronic depression, viral infections, and various types of physical injury is nothing new to herbalists—we have been using this herb in all these capacities for hundreds of years. Recently, though, much attention has focused on Saint John's wort. In Germany, preparations of Saint John's wort are prescribed as the most popular antidepression medicine in clinical practice. Its popularity in Germany triggered an unprecedented research effort worldwide, and many of the studies conclude that Saint John's wort not only can work as well as prescription antidepression drugs, it might work better and be safer. Almost overnight, Saint John's wort became a multimillion-dollar-a-year herb sensation. Recently, many large herb product

Saint John's Wort   *Hypericum perforatum*
Inset: Saint John's Wort leaf   *Hypericum perforatum*

manufacturers have been unsuccessful obtaining enough of the herb to fill demand.

It may be a fabulous herb of commerce and a useful healing herb, but Saint John's wort is listed as a "noxious weed" in most of its North American range. In southern Oregon and northern California, successful eradication campaigns directed at Saint John's wort have been used as celebrated models in "the war on weeds." Saint John's wort generates strong feelings in people who see it as competing with native plant species and degrading the grazing value of public and private rangelands. It makes for poor forage and may cause a photosensitive dermatitis in animals (and people) who consume large amounts of the herb. The consensus has been to kill it, even if that means poisoning the environment. Now, we recognize this weed as having something important to offer us.

Alternatives and Adjuncts: To treat wounds and burns, consider using aloe, comfrey, hound's tongue, or calendula as alternatives or adjuncts to Saint John's wort. For nerve injuries or impairment, investigate the use of skullcap, oatstraw, or bugleweed. For chronic depression, study passion flower *(Passiflora incarnata),* lemon balm *(Mellisa officinalis),* and lavender *(Lavendula officinalis).*

Propagation and Growth Characteristics: Saint John's wort is a perennial that reproduces from short runners or by seed. Some introduced species, such as *H. perforatum,* grow in the same areas where they first became established through livestock grazing decades ago. Wherever this plant takes hold, it becomes a stubborn and persistent resident.

Saint John's wort reproduces freely and vigorously. It ruthlessly competes with native plants and should not be introduced into pristine areas under any circumstances. Unless Saint John's wort already grows around your home, I strongly discourage introducing it into the herb garden. The seeds are tiny and are quickly distributed by birds, wind, rain, and humans.

Gathering Season and General Guidelines: Always wear gloves when gathering this herb. Although reports are rare, some people experience a photosensitive reaction after maximum skin absorption of the active constituents. If you have that reaction, don't use this herb. Also beware of possible herbicides when you gather this plant.

Harvest Saint John's wort according to how you plan to use it. If the focus is on antidepression, gather Saint John's wort when it is just reaching full bloom (late June to August). Use a sharp pair of pruning shears to clip the top 8 to 12 inches from mature plants with lots of flowers. If you want to capture the vulnerary (wound and burn healing) qualities of the plant, wait until the flowers have developed into seed capsules, which contain the greatest concentrations of wound-healing constituents.

Care after Gathering: Saint John's wort should be used or made into medicine shortly after it has been harvested. The herb can be dried for use within a couple of weeks, but many of the active constituents begin breaking down soon after harvest. Saint John's wort makes an oil infusion that is as beautiful as it is useful. The finished product is a clear, ruby red, like claret wine, and is an excellent addition to salves and ointments. It is also good as an oil, for topical application. To use it in oil, first wilt the herb—by reducing the water content, you will lessen the risk of premature spoilage. After the herb has wilted, add just enough alcohol (vodka will work fine) to lightly wet it, then cover it and let it stand for an hour or two before adding the oil. The alcohol solvent breaks down the plant structure and yields an oil of optimum potency. (For more on making oil infusions, see "The Basics of Making Herbal Preparations.")

For internal uses, a tincture is best.

∿Tincture

Fresh, chopped flowers, stems, and leaves: 1:2 ratio in 50 percent alcohol.

Plant-Animal Interdependence: Saint John's wort reputedly is toxic to grazing livestock. Because dense stands of Saint John's wort are usually kept off-limits to cattle, the plants provide sanctuary for wildlife. The flowers attract bees and other pollinators.

Tread Lightly: In recent years, an insect has been introduced into populations of Saint John's wort to provide biological control of this plant. Whether this little bug is eating only Saint John's wort remains to be seen. Take extra care not to disperse either the bugs or the seeds away from established stands. Shake off your herbs well before transporting inside sealed containers. Better yet, tincture it in the field.

> **WARNING! This plant has caused rare cases of photosensitive reactions (severe sunburn) in livestock, particularly animals with light skin. Although toxicity in humans has been limited to a few unconfirmed cases involving ingestion of large quantities by extremely fair-skinned individuals, caution and moderation is advised in its use.**

 # Self-Heal
Prunella vulgaris

Other Names: Heal-all, prunella
Parts Used: All aboveground parts of the flowering plant
Actions: Astringent, tonic, vulnerary, and anti-inflammatory

Self-heal is distinguished from other mints by its unique terminal flowers, which whorl together into a tight, sausage-shaped, purple head. The rest of the plant is a typical mint, except that it has no distinctive, minty fragrance. The leaves are opposite, lance-shaped, and inconspicuously toothed. The stems are four-sided and structurally weak, a trait that often causes the lower portion of the stem to grow horizontally before growing skyward. This plant varies in size according to climate, soil quality, and available moisture, but expect to see most plants in the range of 3 to 12 inches.

Habitat and Range: Self-heal grows just about anywhere—lawns, roadsides, pastures, and subalpine meadows—but it thrives where rich, moist soils have been moderately compacted by livestock, vehicles, or people. It is widespread across North America.

Applications: Historically, self-heal has been used as a medicine for nearly everything. Modern herbalists know it as an excellent topical emollient, astringent, and vulnerary agent. It is an ingredient in several commercial all-purpose salves, ointments, and lotions intended to soothe and speed the healing of minor burns, wounds, and other irritations. For the outdoors en-

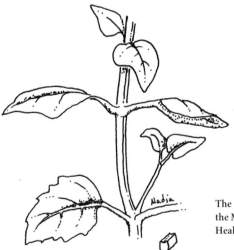

The characteristic square stem of the Mint family is obvious in Self-Heal *(Prunella vulgaris)*.

Self-Heal (Heal-All)   *Prunella vulgaris*

thusiast, the juice of a crushed stem or two will soothe nettle stings, minor bouts with poison ivy, insect bites and stings, and various other trail annoyances. Internally, herbalists use the tea to relieve gastritis and diarrhea, and to aid in the healing of digestive ulceration. Self-heal contains ursolic acid, a compound that has diuretic and antitumor qualities. Researchers are investigating self-heal in cancer studies.

Alternatives and Adjuncts: For use in salves, self-heal combines well with Saint John's wort and chickweed. For treating digestive tract disorders, consider using cleavers, catnip, pineapple weed, chickweed, plantain, shepherd's purse, or uva-ursi as possible alternatives. For specific treatment of stomach ulcers, herbalists sometimes combine juice from self-heal with juice from cleavers.

Propagation and Growth Characteristics: Self-heal is a rhizomatous perennial that can be introduced into the herb garden by transplanting root cuttings. To thrive, the plant needs consistent moisture, rich soil, and usually at least a few hours of shade each day. The plant produces viable seed that can be planted into the garden, but the seeds are tiny and disperse quickly after maturing.

Self-heal grows on the fringes of areas that receive constant soil compression, probably because the plant's shallow rhizomes require heavy soil density to support the weak-stemmed upper plant.

Gathering Season and General Guidelines: Gather this plant while it is blooming. The juiciest, most useful part is the base of the stem, so cut the plant just above ground level with sharp shears. Take care not to pull the plants when cutting or you will damage the roots and compromise next year's growth. Avoid marshy areas and gather during dry weather to prevent excessive soil compression.

When gathering this herb around livestock, be aware that herbicides may be present, particularly if such plants as Saint John's wort, knapweed, or tansy are nearby. Avoid gathering anywhere near roadways, as this plant readily collects lead compounds and other toxic substances.

Care after Gathering: Self-heal is best used fresh, but you can dry and store it for up to six months for use in tea. It has a pleasant flavor and is useful in poultice form for soothing sore gums and minor injuries and irritations to the skin. According to herbalist Michael Moore, fresh self-heal juice can be preserved by mixing it with a small proportion of vodka (75 percent juice, 25 percent vodka). This formula can be stored and used as needed on minor wounds and irritations.

If you dry this herb, give it plenty of air circulation to prevent mold growth. If you make an oil infusion from the fresh herb, let the plants wilt for a full day to increase the shelf life of your oil. (For more on making oil infusions, see "The Basics of Making Herbal Preparations.")

Plant-Animal Interdependence: Although self-heal is not a primary forage plant, most herbivores will eat it. The rhizomes are soil aerators and help allow the introduction of beneficial organisms. The flowers attract pollinators, drawing them down where other plants will benefit, too. Where self-heal grows in abundance on the shoulders of roadways, it helps prevent erosion and collects some of the toxic compounds that otherwise would be washed into adjacent habitats.

Tread Lightly: Don't gather this plant from areas where it serves as a source of forage. There's plenty to go around—find another stand.

 **Shepherd's Purse**
Capsella bursa-pastoris

Mustard Family
Cruciferae

Other Names: Bursa, *Thlaspi bursa-pastoris*
Parts Used: The entire plant, including the roots
Actions: Diuretic, astringent, hemostatic, and emmenagogue

Shepherd's purse is a common lawn and vacant lot weed. It starts out as a basal rosette of petiolate, 1- to 2-inch leaves that are smooth above and hairy underneath. Later, the upper plant consists of one or more slender, erect stems that can grow to 20 inches tall.

The lanceolate leaves of the upper plant grow alternately, clasping the stems at their bases. The lower leaves are deeply lobed; the upper leaves become progressively fewer and less lobed.

Small, white, inconspicuous flowers form on elongated racemes at the top of the plant. The flowers develop into seed-bearing capsules that look like tiny (½ inch or less), sharply heart-shaped purses, a characteristic that earned the plant its common name. These little "purses" are two-celled, have a single ridge along one side, and are slightly concave along the other. Each purse-cell contains three or more tiny seeds.

Many herbalists search only for the "purse" characteristic of shepherd's purse and don't look closely enough when identifying the plant. As a result, many of them gather and use the wrong plant. Although the impostor is usually another harmless member of the Mustard family, it is often a medicinally useless substitute.

Of the plants most commonly mistaken as shepherd's purse, field penny-cress *(Thlaspi arvense)* is at the top of the "Oops" list. Like many other plants mistaken for shepherd's purse, field pennycress seedpods are ovate to nearly circular, not heart-shaped. Like shepherd's purse, the capsules of the look-alikes have two cells, but each cell has only two seeds; shepherd's purse always has several seeds. Farther down the list of impostors is wild candytuft *(Thlaspi fendleri)*, which has seedpods that more closely resemble those of shepherd's purse but again have only two seeds per capsule cell. The pods are narrower, better described as arrow-shaped not heart-shaped. This plant can also be distinguished from shepherd's purse by its waxy, toothed leaves, smaller overall size, and matlike growth characteristic.

Cutaway view of a seed capsule
from Shepherd's Purse exposes
its multiple seeds

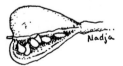

Shepherd's Purse *Capsella bursa-pastoris*

Habitat and Range: Shepherd's purse, a European import, is widely distributed across North America. It is common in cultivated fields, gardens, lawns, vacant lots, areas subject to livestock, and other disturbed areas. It grows in almost any environment, from cracks in city streets to remote mountain campsites. It is adaptable to any elevation, from below sea level to timberline.

Applications: Like many useful wayside weeds, we usually don't think of shepherd's purse when a need arises that it could meet. I love telling people about shepherd's purse, because it reminds us of the possible healing agents that might be as close as some clippings wilting in a bag on the lawn mower.

Herbalists know shepherd's purse as a gentle diuretic that has a special affinity for removing excess uric acid from the body. This makes it useful in the holistic treatment of gout and rheumatoid conditions that are mainly the result of poor liver metabolism and the body's inefficient elimination of postdigestive waste. Shepherd's purse is also a hemostatic—it will help stop bleeding, both internally and externally. As an astringent, it is sometimes used in the symptomatic treatment of diarrhea. Shepherd's purse is said to stimulate menstruation.

Alternatives and Adjuncts: Diuretic or liver supporting alternatives and adjuncts to shepherd's purse include burdock, dandelion, yellow dock, red clover, and Oregon grape. For hemostatic actions, look to yarrow, alumroot, and uva-ursi as stronger alternatives with broader spectrums of usefulness.

Propagation and Growth Characteristics: Shepherd's purse is an annual that seeds itself abundantly. It tolerates varied conditions but prefers rich, well-drained soil and generous amounts of sunshine.

This plant is prolific and becomes a permanent resident if left alone in the garden. It competes well with other "weeds" but generally does not overwhelm them, growing as part of a natural bouquet of grasses and other plants. Nevertheless, many people view shepherd's purse as a "troublesome weed" and subject it to eradication efforts that are subsequently toxic for everyone. For this reason, do not transport this herb into areas where it is not yet established, even the well-kept garden.

Gathering Season and General Guidelines: So long as shepherd's purse looks green and healthy, it is in usable condition. To assist in its proper identification, though, you'd be wise to wait until the plants form seed capsules.

To collect shepherd's purse, all you need is a pair of gloves, a healthy back, and perhaps a hand trowel. Pull the entire plant, roots and all. Keep the plants whole until you use them, to slow oxidation and the loss of medicinal potency.

Care after Gathering: This plant does not keep well and should be used or tinctured the week you harvest it. To use the herb in tea, dry it on butcher paper and use it shortly thereafter.

∾Tincture

Fresh, chopped herb: 1:2 ratio in 50 percent (100 proof) alcohol. *Dried herb:* 1:5 ratio in 50 percent (100 proof) alcohol.

Fresh shepherd's purse is edible and highly nutritious. It has a peppery flavor and is a tasty addition to salads and sandwiches when gathered young. The seeds may be dried and used as a pepper substitute.

Plant-Animal Interdependence: Most herbivores like the taste of shepherd's purse, though it seldom is a primary forage plant. The peppery seeds and seed capsules are consumed by an assortment of birds and small rodents.

Like all annuals, shepherd's purse gives itself back to the soil each year. Unlike most annuals, though, it adds extra nutrients through a specialized process; the seed capsules, when wet, become sticky and attract ants and other insects. Many of these creatures stick to the capsules, die, and decompose in the soil to nourish shepherd's purse seedlings and neighboring plants.

Tread Lightly: Shepherd's purse is an abundant, weedy herb that probably will endure humanity longer than humanity can endure itself. Use it instead of other species that are less resilient to human pressures. Leave the distribution of this herbal ally to the real experts, the plants and animals.

Uva-Ursi

Arctostaphylos species

Heath Family

Ericaceae

> **Other Names:** Kinnickinnick, bear berry, manzanita, Indian tobacco
> **Parts Used:** The leaves, flowers, and berries
> **Actions:** Astringent, antibacterial, and hemostatic

The *Arctostaphylos* genus may be divided into two categories of evergreen shrubs: small, mat-forming, densely branched ground covers with woody, trailing stems *(A. uva-ursi)* and small to medium (up to 10 feet tall), erect hardwood shrubs or trees *(A. manzanita)*. Despite size differences, all species have leaf, flower, and fruit characteristics that distinguish them as members of the *Arctostaphylos* clan. The leathery, alternate leaves are spoon- to lance-shaped, with the upper surfaces darker green than the undersides. The flowers are pink and urn-shaped, arranged in nodding, few-flowered terminal clusters. The fruits are mealy red berries that look like tiny (¼- to ½-inch) apples.

Habitat and Range: Uva-ursi inhabits open forest clearings, from the montane zone up to timberline. *A. uva-ursi* is the predominant circumboreal species, ranging across the northern third of the United States and Canada. That region does not usually support the larger shrub species of *Arctostaphylos,* but in the mountains of Washington, Oregon, Nevada, Arizona, New Mexico, and, especially California (where manzanita is protected by law), several of the larger shrub species predominate.

Applications: Uva-ursi is high in tannin, which makes it a powerful astringent. Internally, it is useful to treat acute urinary tract infections but can irritate the kidneys and bladder if used for more than a few days. Because it's so strong, most herbalists reserve it for acute and severe cases of urinary tract inflammation that require fast-acting intervention. After inflammation subsides, the uva-ursi can be discontinued and replaced with a more gentle alternative. A decoction of uva-ursi is sometimes used as a remedy for acute diarrhea. It can also be useful as an astringent skin rinse for symptomatic treatment of such inflammatory skin afflictions as poison ivy and insect bites. Uva-ursi is sometimes used in a sitz bath to reduce swollen hemorrhoid tissues or postpartum swelling.

Alternatives and Adjuncts: For urinary tract infections, juniper, rose bark, or alumroot are effective substitutes for uva-ursi. Cranesbill geranium, pyrola, goldenrod, couchgrass *(Agropyron repens),* or cranberry juice are weaker, long-term alternatives that are less irritating to the urinary tract and kidneys.

Uva-ursi (and other urinary tract astringents) combine well with such soothing, demulcent herbs as marshmallow *(Althea officinalis),* plantain, or fireweed. These herbs help lubricate and protect the urinary tract during the passage of superfluous materials (such as gravel).

Uva-Ursi   *Arctostaphylos uva-ursi*

Uva-Ursi fruits   *Arctostaphylos uva-ursi*

Propagation and Growth Characteristics: Uva-ursi is a hardy perennial that can be established from transplants of the stem and rhizome cuttings. The seeds must be cold-stratified to germinate. In the wild, the seeds are distributed after passing through the digestive tracts of birds and mammals, which might accelerate the germination period by breaking down the protective, outer shell of the seeds. If you want to propagate from seeds, collect the berries as late in fall as possible and freeze them until you sow them in the spring. Or direct seed during the fall. Either way, they take two or more years to germinate.

You can start root cuttings indoors, in a medium of peat moss and perlite. Keep the cuttings evenly moist until they sprout.

This plant will establish itself only in an environment that fits its habitat needs. Don't try planting uva-ursi into your rock garden if you live in Palm Springs—it won't work. Some nurseries specializing in native ground covers sell established plants. Manzanita, too, is available through nurseries (though it grows very slowly).

Gathering Season and General Guidelines: If the plants are green and appear healthy, gather uva-ursi anytime. Find a large stand and carefully clip a foot or less from loose runners. If you meet resistance when pulling a runner, stop!—that stem has taken root. Whenever possible, gather from plants that are hanging over a boulder or a steep bank. That way you can clip long, free-hanging stems without damaging the root system or the soil surrounding the parent plants.

Although this plant looks as if it would be resilient, I have found that the effects of overharvesting can remain visible for several years. Gather small quantities from multiple sites and monitor the long-term effects of your gathering.

Care after Gathering: This plant's leathery leaf and stem structure render it practically insoluble in water. To use it in infusions, first dry the herb to break down the surface structures. If you plan to use it fresh, you can "pickle" it and make it more soluble; wet the leaves with a small amount of hard spirits (brandy, vodka, or the like) and let the herb stand for a few hours. The alcohol acts as a solvent, breaking down the plant tissues and making the constituents more accessible. Then steep the herb in hot water, for use as a tea. If alcohol is unavailable, fresh uva-ursi can be decocted. (For making decoctions, see "The Basics of Making Herbal Preparations.")

By placing ½ to 1 teaspoon of the tincture into 8 ounces of water, you can make a quick tea that is available anywhere. Many herbalists carry the tincture with them when they travel, as an emergency measure in the event of cystitis.

∾Tincture

Fresh herb: 1:5 ratio in 50 percent (100 proof) alcohol. Uva-ursi does not take to glycerine.

Plant-Animal Interdependence: *Uva-ursi* means "bear berries," and bears are not the only animals that eat the fruit of this plant. The leaves, too, are an important source of midwinter forage. Deer, elk, grouse, and squirrels dig through the snow in search of uva-ursi berries, a rich source of vitamin C and other nutrients. Without hardy evergreens like uva-ursi, those animals might not survive the winter.

The creeping rhizomes make this plant an effective erosion control agent, particularly when it grows on hillsides or at the edge of a steep bank.

Tread Lightly: This plant is abundant and widespread, so there's no excuse for overharvesting it. Minimize your intrusion by gathering conservatively from several different sites. Avoid walking on the plants and never gather when the soil is wet and vulnerable to pressure from human feet.

Because the evergreen plants are attractive in bouquets, several species of *Arctostaphylos* are popular in the floral industry. Manzanita is protected by California state law, but I have witnessed commercial-scale harvesting of this beautiful, slow-growing shrub for the floral industry.

> **WARNING! Because of this plant's rich tannin content, it should not be used internally during pregnancy or in the presence of kidney disease.**

Valerian
Valeriana species

Valerian Family
Valerianaceae

Other Names: Tobacco root

Parts Used: Primarily the fall root; the upper parts of the plant are useful too, but make weaker medicine

Actions: Sedative, antispasmodic, carminative, and hypotensive

Many people say that the strongly aromatic roots of valerian smell like dirty gym socks. I once shared that opinion, but now I relate valerian's odor to something unique and pleasantly earthy. Once you've smelled this plant, identifying it again is as simple as poking your finger into the soil and scratching a root. Valerian foliage is unique, too. The plant emerges as a cluster of loosely arranged lance-shaped leaves, which usually remain larger than the leaves of the mature upper plant. The upper leaves are opposite and pinnately divided, and become progressively smaller toward the top of the plant. The flowers are small, have three stamens (the pollen-bearing part), range from white to pink, and form in branched, terminate clusters. The roots are stringy, brown, and pungent. Several *Valeriana* species grow throughout North America. The main difference between species is size; most of them are less than 1 foot high.

Habitat and Range: Valerian grows in soils that retain moisture well into summer. Look for it on north-facing banks and hillsides or in partially shaded soils high in organic matter. Several native species grow throughout the western third of the United States and Canada. It is less common in the Midwest and East, but in some areas (especially New England), *V. officinalis* has escaped cultivation and is doing well.

Applications: Valerian is the most widely recognized herbal sedative. It works safely and gently to calm the nerves and promote physical relaxation. It does not induce an altered state, as you would expect from a prescription sedative or from alcohol. Herbalists use valerian to relieve insomnia and nervous anxiety, and to help the body relax when it is in pain. Because valerian is a carminative and antispasmodic that has an affinity for the digestive tract, it is useful in circumstances where nervousness is compounded by a spastic colon or an upset stomach. Because of valerian's soaplike saponin content, too much of this herb will have a reverse effect in the stomach, acting as an irritant and possibly an emetic. A small percentage of humans and animals will experience a completely opposite effect from valerian: it will act as a stimulant rather than a sedative. Valerian is very useful for calming animals during thunderstorms or trips to the vet or groomer.

Contrary to what many people believe, valerian was not a precursor to Valium tranquilizers. Perhaps the brand name was inspired by valerian, but the relationship ends there.

Alternatives and Adjuncts: For nervous jitters, skullcap may be a useful adjunct to valerian, or even a better alternative. For anger and rage, herbalists often turn to passion flower *(Passiflora incarnata)*. Hops *(Humulus lupulus)* and oatstraw *(Avena sativa)* are also sedative herbs to consider as adjuncts or alternatives. Depending on the specific symptoms, valerian may be inadvisable in cases of chronic depression, in which case Saint John's wort may be a better alternative. Valerian is sometimes combined with bugleweed (*Lycopus* species) to treat migraine headaches. It is often the herb of choice for people who are trying to break a cigarette or alcohol habit.

Propagation and Growth Characteristics: Valerian is a perennial that can be started from seed or from young transplants. It likes rich soil with moderate amounts of organic matter and plenty of water, though it will survive drought conditions. It tolerates a wide range of pH levels and prefers partial shade to full sun.

Sitka Valerian   *Valeriana sitchensis*   Valerian   *Valeriana officinalis*

Seed collection is often difficult, as the plant dies back quickly after blooming. Mature plants do not transplant well but seedlings do.

Valerian is easy to establish and should be in everyone's herb garden. It is readily available through seed companies and nurseries.

Gathering Season and General Guidelines: With the widespread availability of high-quality cultivated valerian (*V. officinalis*), wildcrafting native species of it are unjustifiable except for emergencies.

To harvest valerian from your garden, dig the fall roots of plants that are at least two years old, after the tops have died back; that is when the roots are strongest. As you harvest, you can divide the roots and replant them for a perpetual supply.

If you need wild valerian for emergency field use, gather it conservatively from healthy stands. If the plants have not yet died back, then use the tops and all, and be sure to replant part of the root. If seeds are on the plant, scatter them in a 3-foot radius of the plant you took and cover them with a thin layer of soil and some leaves or other mulch. Avoid gathering when the ground is wet and vulnerable to soil compaction.

Care after Gathering: This herb can be used in fresh or dried form (follow standard drying procedures). The roots' potency increases as they dry, and they will keep for a year or more if properly stored. Use this herb as a tea or tincture.

∿Tincture

Fresh herb: 1:2 ratio in 50 to 70 percent alcohol. *Dried herb:* 1:5 ratio in 50 to 70 percent alcohol.

Plant-Animal Interdependence: Valerian is browsed by virtually everything in its habitat. I believe animals use the plant for more than food; from what I have observed, wild species of valerian are only eaten in measured quantities, once in a while. My guess is that this early spring plant is a digestive aid for animals that have been foraging on bark, twigs, and other fibrous food sources all winter.

Tread Lightly: We cannot justify harvesting wild valerian. In some areas of the western United States, populations of Sitka valerian (*V. sitchensis*) have been seriously compromised by commercial wildcrafters. If you purchase commercial preparations of valerian, please ascertain that it comes from a legitimate cultivated source.

Wild Ginger
Asarum caudatum

Birthwort Family
Aristolochiaceae

Other Names: Canada snakeroot

Parts Used: All parts of the plant are useful, with potency relative to aromatic strength

Actions: Vasodilator, tonic, carminative, rubifacient, and stimulant

Wild ginger is completely unrelated to grocery store gingerroot, but it smells nearly identical. The slightest disturbance of wild ginger foliage brings a burst of gingerlike scent guaranteed to delight. Wild ginger is a perennial herb that reproduces mainly from its extensive rhizomes, a characteristic leading to dense colonies of the plants carpeting the forest floor. The 2- to 5-inch-wide, dark green, somewhat hairy leaves are broadly heart-shaped (nearly circular) and form long (up to 8-inch) petioles that extend directly from the rootstock. The brownish purple, 1- to 3-inch-wide flowers are well described as "odd," with three lobes tapering and curling away from the rest of the flower like

Wild Ginger *Asarum caudatum*

Wild Ginger 〜 203

insect feelers. Each solitary flower forms the leaf axils, at ground level, making them almost invisible beneath the plant's large leaves. *A. caudatum* is the most widespread species. All the *Asarum*s share a similar appearance, but in my experience species in the West are more aromatic than their eastern relatives.

Habitat and Range: Wild ginger can survive in disturbed areas, but it won't do well there. It is a plant of the dark forest, requiring thick compost and deep shade to flourish. It typically grows in old-growth forests, where it is important in aerating easily compressed soils. In the West, wild ginger inhabits mountains and rain forests from British Columbia south through similar habitats in Washington, Oregon, northern California, and the Idaho Panhandle. A variation of the same species grows from the eastern Canada provinces south to North Carolina. I have seen it growing in abundance in the hardwood hollows of central and northern Appalachia. Other, less common species of *Asarum* grow sporadically in their eastern and western ranges.

Applications: Wild ginger is a secretory stimulant and peripheral vasodilator that helps to improve venous blood return at skin level. Like its culinary counterpart, wild ginger promotes relief from indigestion. Because its usefulness parallels culinary ginger, how can we justify the wild harvest of this plant (except for an urgent need in the field)? The greatest healing gift offered by wild ginger comes from recognizing and appreciating what a healthy stand of these plants represents—an intact ecosystem.

Alternatives and Adjuncts: The culinary variety of gingerroot *(Zingiber officinale)* serves as a more effective and earth-conscious substitute for wild ginger. For indigestion and other gastric disorders, chamomile, pineapple weed, catnip, peppermint, or fennel *(Foeniculum vulgare)* could be good substitutes. When a peripheral vasodilator is indicated, cayenne *(Capsicum minimum)*, yarrow, and *Ginkgo biloba* are alternatives.

Propagation and Growth Characteristics: Wild ginger, a perennial, requires deep shade and consistently moist, acidic soil rich in forest compost. If you replicate that habitat as accurately as possible, you can cultivate wild ginger from carefully handled rhizome cuttings.

Wild ginger typically grows in dense patches that carpet the forest floor. What looks like several individual plants may be offshoots from a single rhizome. The rhizomes require a deep accumulation of forest debris, and the plant does not tolerate a full day of sun.

Gathering Season and General Guidelines: If you collect wild ginger, try to gather from plants that have already flowered and dropped their seeds. From my experience in the Pacific Northwest, the leaves typically are the most potent part of the plant, but this may vary with habitat and growth characteristics between plant communities. Selectively clip off one leaf from each of several plants. If you insist on using the rhizomes, always replant a piece of it in the hole you created, and cover it with ample forest compost. Stay on established paths and trails when the forest floor is wet. Gather from the periphery of healthy stands, taking care not to compress the soil and compost above the plants' horizontal root system. Best of all, use an alternative herbal medicine.

Care after Gathering: Like most herbs from shady, damp environments, wild ginger can mold quickly. Once the roots and leaves dry, they lose much of their pungency and usefulness, so harvest only what will meet your immediate needs and make a tincture using the fresh herb.

❧Tincture
Fresh herb: 1:2 ratio in 50 percent alcohol.

Plant-Animal Interdependence: Wild ginger's extensive, horizontal root system forms an integral life-support structure for subterranean organisms beneath the forest floor. Wild ginger's neighborhood usually comprises thick, acidic forest floor debris that composts slowly and gradually compresses under its own weight. This thick mat can become totally impervious to water, microorganisms, heat, and even air. Add to this the minimal sunlight typical in old-growth forests, which results in very little photosynthesis, and you have a habitat of specially adapted organisms. Wild ginger is a key element in the forest floor biocommunities, maintaining the delicate balance between life and sterility.

Tread Lightly: When you visit old-growth forests, spend some time looking around. Meet the elder plants and their offspring, and pay attention to what they teach you. Wild ginger is still abundant in North America, but its delicate and pristine habitat is rapidly disappearing. Because human feet can do more damage than overharvesting in the ancient rain forests where *Asarum* grows, please refrain from harvesting this plant.

Willow
Salix species

Willow Family
Salicaceae

Other Names: Pussy willow, red willow, weeping willow
Parts Used: The bark, and to lesser effect, the leaves
Actions: Analgesic, anti-inflammatory, antipyretic, astringent, and antiseptic

Salix is a huge, widely varied genus of deciduous shrubs and trees. Describing the variables (and the hybridization possibilities) among the one hundred or more species in temperate North America would be impossible here. As an introduction to the genus, though, we can examine its most typical characteristics. All willows have a unique flower (a catkin) and distinctive leaf buds. The catkins form along young twigs and side branches in early spring, male and female catkins on separate plants. The catkins first appear as silvery white fluffs of fur, what we commonly know as pussy willows. This stage is brief; they soon become more like a bristly, woolly worm as they develop toward sexual maturity. The leaf buds are sheathed by a single, smooth, brownish scale, like the covering on a kernel of popcorn. The scale stays attached until the catkins have fallen off and the leaves are nearly mature. The young leaf shoots have a small, winglike appendage at the base.

Habitat and Range: Willows typically grow near water or in soil that holds moisture over long periods. *Salix* habitat varies widely from low desert to alpine elevations across North America. A field guide for your locale will help you identify willows—look for them along rivers and streams.

Applications: Willows contain salicin, a salicylic acid–bearing glycoside that acts as an analgesic. Salicylic acid was a precursor to aspirin. Herbalists use willow bark as a plain and simple alternative to aspirin. Many people don't know that willow bark can also be used in topical applications to relieve the surface pain and swelling of welts, burns, cold sores, and other irritations. Unlike aspirin, willow adds astringency, making it an anti-inflammatory pain relief option for acute irritations of the mouth and gastrointestinal tract. Like aspirin, though, willow may irritate stomach ulcers and other disorders where aspirin is specifically contraindicated. Willow will not thin the blood, as aspirin does, so it is not an effective alternative for preventing cardiovascular disease.

I always have willow in mind as a widely available, impact resilient pain reliever if I don't have an aspirin. But, in my experience, a headache will usually go away before one can ingest enough of this herb to be of remedial value. Further, willow bark is not 100 percent reliable in its painkilling activity. Because the salicilate content of willow varies from plant to plant, I can't help but weigh the advantage of harvesting a plant that is critical to its environment against a better alternative contained in a little white tablet.

Willow catkins   *Salix* species

Geyer Willow   *Salix geyeriana*

Willow leaf bud   *Salix* species

Alternatives and Adjuncts: Meadowsweet *(Filipendula ulmaria), Spirea* species, or quaking aspen *(Populus tremuloides)* are all more consistent than willow bark in their anti-inflammatory and painkilling capabilities, and therefore are better choices than willow when an herbal analgesic is indicated.

Propagation and Growth Characteristics: Willows are persistent plants. They reproduce from root runners or by seed, and many species can be propagated simply by placing cuttings of the young branches into a glass of water until they take root. They are fast growers—many of the erect, shrub species, such as *S. exiqua* or *S. scouleriana,* can grow back to their original height just two years after being cut down to the ground. I once used some young, straight willow branches for pea stakes, and when I pulled them up in fall, they had taken root! Several species of *Salix* are available through nurseries.

Gathering Season and General Guidelines: The best time to gather willow bark is in spring, when the catkins are gone and the leaf buds are just beginning to open. That is when the cambium (inner bark) is transferring the heaviest concentration of water, nutrients, and medicinal constituents from the roots to the leaves. The bark will be easy to strip off with a small pocketknife. Do not cut completely around the circumference of a limb or you will kill it, and limit your cut to no more than one-third of the circumference. A better method is to clip a few limb ends and either shave the bark off with your pocketknife or cut them into small pieces for making tea. You want the greenish inner bark (the cambium), not the tough inner tissue (wood). Cut conservatively from several trees and monitor the effects of your cutting over time to be sure you are not interfering with the trees' growth.

Care after Gathering: You can use the bark fresh or dried in decoctions, but it is always better used fresh. Dry the bark in an open paper bag or on butcher paper. Once it is fully dried, store it for use in teas or decoctions; it should keep through the year.

ᦉTincture
Fresh bark: 1:2 ratio in 50 percent alcohol.

Plant-Animal Interdependence: Virtually everything in the forest eats willow. In spring, deer, elk, and rodents go after the young leaf buds, and bees relish one of the early sources of nectar. Animals browse willow leaves all summer long and strip the bark in fall for its energy-rich carbohydrates. Maybe animals use this plant for pain relief, too.

Willows are strong and resilient, providing a safe haven for insects, rodents, and nesting birds. Shrub species collect debris in the lower branches, creating specialized habitat for a wide variety of creatures.

When this valuable food source is heavily foraged, it becomes inedible until a balance is restored. We don't know exactly how or why this happens, but it contributes to cycles of natural die-off that keep animal populations in check. The plant may produce compounds that foraging animals find unpalatable.

Tread Lightly: Although this herb is profusely abundant, its roles in nature are critical. If the willows in your area are being devoured or certain animal populations, such as rabbits, are rapidly increasing, ease the plant's burden by seeking an alternative or by doing without.

Yarrow
Achillea millefolium

Sunflower Family
Compositae

Other Names: White yarrow, milfoil, soldier's wound wort, thousand leaf, field hop, sneezeweed

Parts Used: All aboveground parts of flowering plant

Actions: Antiseptic, hemostatic, tonic, hypotensive, astringent, diaphoretic, diuretic, and insect repellent

Yarrow has flat-topped terminal clusters of small white to pinkish white flowers and alternate, finely dissected, feathery leaves (*millefolium* means "thousand-leafed"). The entire plant is strongly aromatic, with a pungency similar to mothballs. The stems are typically woolly-hairy. Several yarrow cultivars have been developed for the floral industry, and others have escaped cultivation in many areas. Those plants are not as cold hardy or as pungent as their wild cousins, and bear yellow-, red-, or peach-colored flowers.

Habitat and Range: Yarrow ranges across the Northern Hemisphere, from sea level to above timberline. The density of plant populations increases to the north from the central latitudes of North America.

Applications: Yarrow is abundant and exceptionally useful. Herbalists know it as one of the best herbal antiseptic and hemostatic remedies. It has been used traditionally to stop bleeding and prevent infection in open wounds on the battlefield. Though the hemostatic qualities of yarrow stop internal or external bleeding, it does not act as a direct coagulant in the bloodstream; surprisingly, it works as an excellent vascular tonic that helps improve circulation. When taken internally, yarrow tightens capillary walls, making them stronger while at the same time initiating a vasodilating action that improves circulatory capacity, especially in the smaller, peripheral capillaries of the lower extremities. Because of these unique properties, herbalists frequently use yarrow to treat varicose veins and other forms of vascular disease. When yarrow oil or tea is applied externally, it has a reverse effect, acting as a vasoconstrictor on subcutaneous capillaries. This makes the herb especially useful for checking the bleeding of tiny, broken blood vessels that make an unsightly mess just beneath the skin.

Because yarrow has such a pronounced peripheral vasodilating effect, the tea tends to warm the skin, flushing it pink and triggering a diaphoretic (sweating) response. Herbalists rely on this action to help draw the heat of an illness outward to the surface of the body. From a holistic perspective, sweating is viewed not only as the body's cooling-off mechanism but also a backup for eliminating waste materials that have built up in the outer tissues. The vasodilating activity that flushes the skin also improves lymphatic circulation, boosting the body's effectiveness in "washing" tissues and reinforcing its resistance to infection.

Although it should not be used during pregnancy because of the vascular activities of its volatile oil constituents, yarrow is otherwise very safe. It's a good first herb in the home apothecary of the beginning herbalist.

Fresh or dried, yarrow repels moths.

Alternatives and Adjuncts: Other vascular tonics to use as adjuncts with yarrow or a substitute for it are ginger (the culinary type), elderberry, cayenne, horse chestnut *(Aesculus hippocastanum)*, garlic, and hawthorn. For use in cold-formula teas, yarrow combines well with arrowleaf balsamroot, echinacea, elderberry, culinary ginger, coltsfoot, mullein, or grindelia *(Grindelia squarrosa)*.

Propagation and Growth Characteristics: Yarrow is a rhizomatous perennial that reproduces readily from tiny seeds. Although many people regard it as a "troublesome weed," it is not as competitive as it is prolific and

Common Yarrow   *Achillea millefolium*   Yarrow leaf   *Achillea millefolium*

does not compromise neighboring plants. If you choose to introduce this plant into the herb garden, remember: each plant will produce thousands of tiny seeds that will make it a long-term resident.

Young plants transplant easily, or you can sow the seeds directly into the garden. To gather seeds, wait until the flowers are beginning to dry, then clip off one or two and put them in a paper bag to dry completely. Don't pull your hair out trying to separate the seeds from the flower petals—dump the whole mess where you want yarrow to grow.

Yarrow has a long and prolific bloom period. Flowers usually appear in midspring and last through August. Clipping the top flowers encourages side shoots and more flowers.

Gathering Season and General Guidelines: Gather yarrow anytime during its bloom period. Using sharp clippers, remove a few flower heads from several plants, leaving plenty for reseeding and pollinators. The entire plant is useful, but flowers make the best tea. If you plan to make a tincture, cut mature plants at least 6 inches above ground level. By using the leaves and all, you can concentrate your harvest and avoid trampling on surrounding vegetation.

Beware of the possible presence of herbicides when wildcrafting yarrow, particularly where it is growing in cultivated areas among other "nuisance weeds."

Care after Gathering: Like many herbs, yarrow is more water soluble and better tasting after it has dried. The flowers dry quickly and should be stored in an airtight container right away, as they will literally turn to dust if left unattended. Properly stored, the dried herb will last for about a year.

∽Tincture
Fresh herb: 1:2 ratio in 50 percent alcohol. *Dried herb:* 1:5 ratio in 50 percent alcohol.

Plant-Animal Interdependence: Yarrow is seldom foraged, but it serves dual roles in insect-plant interdependency. The pungency of this plant repels many insects but attracts others. This selective function provides a balance of biodiversity that benefits yarrow and its neighbors. The sudden removal of this plant from its biocommunity may have ill effects on companion plants and other organisms.

Tread Lightly: Before gathering this herb, take time to review its natural roles. Look closely to see what may be living in and around these plants.

Yellow Dock
Rumex crispus

Other Names: Curly dock, sour dock, sour grass
Parts Used: The root
Actions: Alterative, cholagogue, and laxative

Yellow dock is a hearty, taprooted perennial that can grow 5 feet tall. The elongated (up to 12 inches long and 4 inches wide), lance-shaped, basal leaves form on proportionately long leaf stems (petioles) and often curl at the margins. The alternate stem leaves are smaller but more numerous. The single, stout stem is commonly red and bears long, terminate clusters of small, greenish white flowers above the rest of the plant. As the flowers mature and dry, they turn rusty red and contrast with the surrounding flora. Of the twenty-

Yellow Dock   *Rumex crispus*   Sheep Sorrel   *Rumex acetosella*

five or more *Rumex* species that inhabit North America, *R. crispus* typifies the larger ones. At the smaller end of the genus, such species as sheep sorrel *(R. acetosella)* look different and vary in usefulness.

Habitat and Range: Yellow dock is widely distributed in disturbed areas across North America. The *Rumex* genus is an import from Europe.

Applications: Herbalists traditionally use yellow dock as a liver and digestive "cleansing herb." It eases psoriasis and other skin conditions that liver dysfunction promotes. Because yellow dock is high in iron, it is sometimes used to treat anemia. Its laxative properties make it a useful remedy for constipation.

Yellow dock is a primary ingredient in many anticancer herbal formulas, in the belief that it helps rid the body of waste materials that otherwise would stress further an already stressed-out body. It accomplishes this through the combined activities as a liver and gallbladder stimulant, and as a relatively mild laxative.

Alternatives and Adjuncts: For use in liver metabolism disorders, yellow dock will combine with burdock, dandelion root, marshmallow *(Althea officinalis)*, and red clover. For chronic constipation, Oregon grape is a good alternative. For iron-poor blood, such nutritive herbs as nettle are better options.

**Yellow Dock seed
head and upper leaf**

Propagation and Growth Characteristics: Yellow Dock is a perennial weed in most areas but may grow as an annual or biennial in extreme climates. It reproduces readily from seed, and is easy to establish in the herb garden. It will grow in just about any soil but prefers rich, deep loam. It likes moist conditions, but will endure long periods of drought once it is established. Yellow dock is a classic vacant lot weed.

Gathering Season and General Guidelines: Gather the roots in fall, after the plant has gone to seed—that is when they are most potent. The young, green leaves of this plant are edible. Some people prefer yellow dock's robust, tangy flavor to mustard or spinach, but for most people, it's too strong. If you want to eat yellow dock, gather the youngest leaves possible in the spring. The plant (especially older plants) contains oxalic acid, which can be toxic if ingested in large enough quantities.

If not a target itself, yellow dock commonly grows among plants that are treated with herbicides. Avoid gathering this plant from ditch banks or areas that are actively cultivated. Always check for local weed abatement programs before gathering this herb.

Care after Gathering: You can cut the roots into chunks and dry them for use in decoctions or weak infusions. Or cut the roots into three or four pieces and spread them out on paper to dry. Then cut or grind them into smaller pieces when you are ready to use them. Properly stored, they should keep for a year or more.

∽Tincture
Fresh root: 1:2 ratio in 50 percent (100 proof) alcohol.

Plant-Animal Interdependence: Yellow dock is typically the first plant to grow in a heavily impacted area, such as a construction site or plowed field. Its long taproot effectively penetrates and aerates compacted soils and helps prevent erosion in areas where water runoff could be a problem. Like all perennials, yellow dock recycles its body into the soil every year; it is particularly high in iron, a nutrient important to plants as well as people.

Tread Lightly: I enjoy using this plant and teaching people about it because it is a common weed. By educating people to the fact that many plants generally regarded as "just weeds" are medicinally useful, they gain a deeper perspective on the values all around them, which I hope will translate to more weed use and less herbicide use.

> **WARNING! The leaves of this plant contains oxalic acid, a compound that may cause digestive upset if ingested in large quantities. Excessive, long-term use may result in the formation of urinary tract stones. This plant should not be used during pregnancy. The root contains anthraquinone constituents that may cause intestinal cramping and diarrhea if ingested in excessive quantities.**

Herbal Emissary—Naomi A. Pelky

RESOURCE GUIDE

This resource guide lists sources of more information on plant preservation, schools of herbal studies, and suppliers of plants, seeds, herbal products, books, and medicine-making supplies. If any addresses or phone numbers have changed since this book was printed, check the Rocky Mountain Herbalist Coalition's *Direct Marketing Registry* (see "Bibliography and Recommended Reading"), the World Wide Web, or your local herb store for newer information.

About United Plant Savers

The Founding Vision

When I first moved to New England, I was delighted to discover that the Northeast was home to many of the famous medicinal herbs that I loved and used often in my practice. However, I was quickly struck by how many of these plants that used to be plentiful were now threatened or endangered. Having traveled in Europe and other parts of the world where herbal medicine has been highly regarded for centuries, I was alarmed by the crisis in our North American wild herb populations. In most parts of the Western world, there are few stands of herbs growing wild, and many countries, having depleted their own resources, depend on the wild resources of North America to supply their medicinal plant species.

These same very disturbing thoughts inspired me to start a replanting project on our land. Working quietly, often alone, in the backwoods of our 500-acre preserve, I began to slowly replant native species that once thrived in the rich hardwood forests. I felt like I was doing "soul work" as my fingers parted the soil and planted the tiny rootlets of species that once knew this as their home. Out there in the woods, with my simple tools and a bag of roots and seeds, a vision began to unfold. I started handing out roots and seeds to my students, who planted wild gardens in their neighborhoods. Many other herbalists are doing the same thing, following in the footsteps of Elzeard Bouffier, the gentle spirit who reclaimed an entire section of France by planting thousands of trees in his lifetime. Our hope is that once again our woods will be plentiful with the medicines of the earth.

—Rosemary Gladstar,
Founder of United Plant Savers

The United Plant Savers Commitment

Indiscriminate wild harvesting, deforestation, and urbanization have devastated many formerly abundant herb populations. Perhaps even more disturbing, native North American medicinal plants are being exported to meet the demand in other countries where wild plants have already been gravely depleted.

United Plant Savers (UpS) was formed in a spirit of hope, as a group of herbalists committed to protecting and replanting threatened species and to raising public awareness of the plight of our wild medicinal plants. The UpS membership reflects the great diversity of American herbalism and includes wildcrafters, seed collectors, manufacturers, growers, botanists, practitioners, medicine makers, educators, and plant lovers from all walks of life.

UpS Goals

- ∿ Identify and compile information on threatened medicinal plants in each state and/or bioregion
- ∿ Make this information accessible to herbal organizations, communities, and individuals
- ∿ Provide resources for obtaining seeds, roots, and plants for replanting and restoration
- ∿ Secure land trusts for the preservation of diversity and seed stock for future propagation efforts
- ∿ Raise public awareness about the tragic effects of overharvesting, deforestation, and urbanization and the current plight of native wild herbs
- ∿ Identify and disseminate information on therapeutic alternatives to threatened species
- ∿ Encourage more widespread cultivation of endangered medicinal plants and greater use of cultivated plants
- ∿ Develop programs for schools and communities to replant threatened plant species into their native habitats

Join United Plant Savers!

To join UpS or for more information, call or write:

United Plant Savers
P.O. Box 98
East Barre, VT 05649
(802) 479-9825
fax: (802) 476-3722
e-mail: JCATSage@plainfield.bypass.com
www.plantsavers.org

When sending in your membership donation, please provide your name, address (city, state, zip), and your phone and fax numbers. Also include any ideas you have that may help the UpS effort, and list any skills you have to offer. UpS is a nonprofit organization and is eligible for tax-free donations.

Herbal Education

The following herbal schools and individuals provide education focused on the earth-conscious study, harvest, and medicinal uses of plants.

~ California School of Herbal Studies
James Green, director
9309 Hwy. 116, P.O. Box 39
Forestville, CA 95436
(707) 887-7457
www.CSHS.com

The oldest school of herbal studies in the United States, the California School of Herbal Studies offers a comprehensive curriculum that ranges from half-day workshops for beginners to intensive year-long courses of study. Director and cofounder James Green is author of the popular *Male Herbal*, and is a leading educator and visionary in the field of Western herbalism. The faculty of this school includes many of the world's best-known herbalists, and the fundamental theme centers on a deep level of respect and concern for the plants they teach about.

~ Columbines and Wizardry Herbs Herbal Apprenticeship
Howie Brounstein, primary instructor
P.O. Box 50532
Eugene, OR 97405
(541) 687-7114
e-mail: howieb@teleport.com
www.teleport.com/~howieb/howie.html

"It's easy to harvest wild plants, the hard part is not harvesting."
—Howie Brounstein

Howie provides a six-month intensive course in plant identification, ethical wildcrafting, basic herbalism, and wild foods.

~ Friends of the Trees Society
Michael Pilarski
P.O. Box 4469
Bellingham, WA 98227
(360) 738-4972
www.geocities.com/rainforest/4663

Michael Pilarski is a wildcrafter of medicinal plants that grow in the Pacific Northwest. Michael knows a great deal about the medicinal plants of his bioregion, and he performs his craft from a deep-ecological perspective that takes into account the sustainability and interdependent needs of the plants he harvests. Michael offers classes and participates in seminars throughout the year.

> ∾ Northeast School of Botanical Medicine
> 7Song, AHG, director
> P.O. Box 6626
> Ithaca, NY 14851
> (607) 564-1023

7Song and his staff provide beginning through advanced levels of education in herbal medicine, plant identification, and ethical wildcrafting, with emphasis on the sustainability of wild botanicals.

> ∾ Rocky Mountain Center for Botanical Studies
> Feather Jones, director
> P.O. Box 19254
> Boulder, CO 80308
> (303) 442-6861
> e-mail: rcmbs@indra.com
> www.herbschool.com

For decades, Feather Jones has been a strong, positive voice for the continued survival of medicinal plants. The school she founded now carries her caring voice deep into the consciousness of Western herbalism. Located in North America's "hotbed city" of cutting-edge herbal research and education, Rocky Mountain Center for Botanical Studies offers a diverse curriculum of herbal studies featuring many of the best educators in the world.

> ∾ Island Herbs
> Ryan Drum, Ph.D.
> Box 25
> Waldron Island, WA 98297
> (360) 739-4093

Ryan is one of the best-known teachers of wildcrafting herbalists in North America. He has been harvesting and studying wild medicinal plants for decades and offers intensive, hands-on apprenticeships to those with a strong will and perseverance to work hard and learn from someone who knows his craft—and the plants—like he knows himself.

> ∾ Sage Mountain Herbal Retreat Center and Botanical Sanctuary
> Rosemary Gladstar, director
> P.O. Box 420
> East Barre, VT 05649
> (802) 479-9825
> fax: (802) 476-3722

Rosemary Gladstar, author of *Herbal Healing for Women* and the foreword of this book, is one of the best known and most respected teachers of herbalists in the world. She is the founder of United Plant Savers, a nonprofit organization dedicated to saving endangered and threatened medicinal plants, and cofounder of the California School of Herbal Studies. Rosemary offers a full curriculum of seminars and study courses on herbal medicine, plant preservation, and other holistic studies throughout the year. She also offers an excellent home study course.

   ~ Southwest School of Botanical Medicine
      Michael Moore, director
      401 Purdy Lane
      Bisbee, AZ 85603
      (520) 432-5855

Known as a leading expert in the field of botanical medicines of western North America, Michael offers a very comprehensive, four-month course in clinical herbalism. Michael has written several herbal classics, including *Medicinal Plants of the Mountain West, Medicinal Plants of the Desert and Canyon West, Los Remedios,* and my personal favorite, *Medicinal Plants of the Pacific West.* Michael's school is for people who seriously want to learn about the practical, clinic use of herbal medicines, but from a perspective that fully considers the nature of the plants themselves.

   ~ Sweetgrass School of Herbalism
      Robyn Klein, director
      6101 Shadow Circle Drive
      Bozeman, MT 59715-8384
      (406) 585-8006
      e-mail: rrr@wtp.net

Robyn offers an excellent curriculum to all levels of herbal interest. She founded the school in 1996 "from the dream to teach the medicinal uses of plants so that we may better appreciate and sustain them for our children's children."

   ~ Therapeutic Herbalism
      David Hoffmann, B.Sc., M.N.I.M.H.
      2068 Ludwig Road
      Santa Rosa, CA 95407
      (707) 537-9830

Therapeutic Herbalism is perhaps the most comprehensive home study course on herbal medicine available. David Hoffmann, the course author and tutor, is a well-known and highly respected herbalist. A member of Britain's National Institute of Medical Herbalists since 1979, David has written more than eleven books on herbs and herbalism. He is a board member with United Plant Savers and actively promotes and teaches about the deep ecology and preservation of medicinal plants worldwide.

Sources of Threatened or Endangered Plants and Seeds, and Information about How to Grow Them

There are many excellent sources of herb seeds, roots, and plant starts. Here a few of the ones I consider the very best.

      Abundant Life Seed Foundation
      Box 772
      Port Townsend, WA 98368
      (360) 385-5660
      fax: (360) 385-7455
      e-mail: abundant@olypen.com

The goal of this nonprofit organization is to propagate and preserve the genetic diversity of native and naturalized plants. They have been doing this for more than twenty years, and as a result, they offer a great selection of hard-to-find medicinal plant seeds, including many rare and endangered species.

      Horizon Herbs
      Richo Cech, herbalist and owner
      P.O. Box 69
      Williams, OR 97544
      (541) 846-6704
      e-mail: herbseed@chatlink.com

Richo provides an extensive variety of hard-to-find medicinal plant seeds and rootstock, complete with practical information about propagation. Richo's impressive catalog is perhaps the most comprehensive, at-a-glance guide to growing medicinal plants I have ever seen. Richo has also written a complete line of informative booklets about growing and processing many of the species he provides. Besides stocking and distributing medicinal plant seeds, Richo is also a leading expert in the fields of medicinal plant horticulture and wild plant restoration.

      Land Reformers
      35703 Loop Road
      Rutland, OH 45775

An impressive catalog of plants that are indigenous or adaptable to the hardwood forests of eastern and central North America. The "botany boys" who operate Land Reformers know what they are doing, and they sell premium-quality stock.

 ∾ The North Carolina Ginseng and Goldenseal Company
   Robert Eidus, herbalist and owner
   148 Anderson Branch Road
   Marshall, NC 28753
   (704) 649-3536

Aside from growing and providing high-quality, fungicide-free American ginseng and goldenseal roots for transplant or personal use, Robert is leading an unprecedented rescue, relocation, and rehabilitation effort to save endangered wild plants in his bioregion. Robert knows ginseng and goldenseal inside and out. He publishes and distributes affordable, easy-to-use booklets about how to propagate each of these finicky plants.

Retail Sources of Dried Herbs, Herbal Extracts, Medicine-Making Supplies, and Books

 ∾ American Botanical Council
   P.O. Box 201660
   Austin, TX 78720-1660
   (800) 373-7105
   e-mail: custserv@herbalgram.org

Publishers of *Herbalgram* magazine. ABC has perhaps the most comprehensive retail catalog of herb-related books in the world.

 ∾ Meadowsweet Herbs
   P.O. Box 1516
   Missoula, MT 59806
   (406) 728-0543
   e-mail: msweet@bigsky.net

Meadowsweet Herbs offers a full line of herbal tinctures.

 ∾ Mountain Rose Herbs
   20818 High Street
   North San Juan, CA 95960
   (800) 879-3337
   http://www.botanical.com/mtrose/

Mountain Rose offers a retail catalog of herbs, herbal products, books, and medicine-making supplies.

~ Montana Arnica
   P.O. Box 350057
   Grantsdale, MT 59835
   (406) 363-3716

Bulk, fresh or dried organically grown herbs.

~ PetSage
   4313 Wheeler Avenue
   Alexandria, VA 22304
   (800) PET-HLTH (738-4584)
   fax: (703) 823-9714
   e-mail: info@petsage.com
   www.petsage.com

PetSage offers a big catalog of natural products for pets, including a full line of herb glycerite tinctures (formulated by yours truly).

GLOSSARY

action. See *medicinal action.*

adaptogen. A nontoxic substance believed to generally increase stamina and overall energy levels in the body, especially in stressful conditions.

aerial. Aboveground portion of a plant.

allopathy. The use of drugs or other means to antidote a disease or symptom, in a manner not necessarily cooperative with the body's natural functions. Antonym: *homeopathy.*

alluvial. The type of process where sedimentary materials, such as soil and rocks, are deposited or accumulated by flowing water.

alterative. An action that gradually alters an existing condition in the body. A "blood alterative" is often referred to as a "blood cleanser," as it alters the entrance of toxins and waste materials into the bloodstream, in most cases through stimulation of liver function.

alternate. In reference to leaves, alternate leaves are arranged along a stem at various distances from each another, but never opposite each other across the stem.

analgesic. A pain-relieving substance.

anesthetic. A substance that reduces painful sensitivity. Unlike general analgesics, anesthetics often can be applied locally. An injection of Novocain at the dentist is a local anesthetic.

annual. A plant that blooms, distributes its seeds, and then dies during its first and only year. Annuals depend solely on seed reproduction.

anther. The pollen-bearing organ at the end of the stamen, responsible for the distribution of pollen. The anther is usually yellow or orange and is the part bees seek out.

anthropocentric. Regarding the human being as the central fact or final aim of the universe.

anticatarrhal. A substance capable of assisting the body in eliminating excess mucus from the upper respiratory tract, through anti-inflammatory actions on the mucous membranes that are responsible for the secretions.

antifungal. Capable of preventing or inhibiting fungal infections.

antihidrotic. Capable of preventing or inhibiting perspiration.

antihistamine. An herb or drug that alters histamine responses in the body, with the effect of reducing the discomforts of allergic reactions and, in some cases, motion sickness.

antilithic. Capable of preventing the formation of, or aiding in the elimination of, gravel or stones in the urinary system.

antimicrobial. An action that helps the body to resist, inhibit, or destroy pathogenic microbes. In holistic medicine, this term generally describes actions that assist the body in fighting bacteria, fungi, or viruses at their original point of infection.

antioxidant. A general, perhaps worn-out term referring to the ability of a substance to control or eliminate free radicals or reduce cellular oxidation in the body.

antipyretic. Capable of reducing fever.

antirheumatic. Capable of relieving the symptoms of rheumatic conditions, such as rheumatoid arthritis.

antiscorbutic. A substance that helps to prevent or cure scurvy or other imbalances relative to vitamin C deficiency.

antiseptic. Generally refers to substances that kill or inhibit the growth of pathogenic microbes. In holistic medicine, this term describes substances that interfere with bacterial infections regardless of the body's natural abilities to do so. In this context, antiseptic herbs are applied as allopathic remedies.

antispasmodic. Capable of relieving spasms.

antitussive. Capable of suppressing coughing.

antiviral. Capable of inhibiting the reproduction or activity of a virus.

astringent. Capable of tightening soft tissues of the body. Astringents are used to stop bleeding, to reduce inflammation, and to stop diarrhea.

axil. In plants, *axil* usually refers to the junction where a petiole or peduncle joins the stem. Many members of the Mint family, for example, present their flowers at the leaf axils. These are referred to as axillary flowers. The flowers of licorice (*Glycyrrhiza* species) are presented on peduncles (pedicels), which are also borne from the leaf axils.

bacteriostatic. Specifically acting to inhibit the multiplication of bacteria.

basal. Refers to the base of something. Basal leaves are the ones at the extreme lower end of a plant. They are usually the first true leaves to appear after germination.

biennial. A plant that blooms only during its second year of growth, and then dies.

bioregion. A geographical region that is unique in its makeup of resident plants and other organisms. Also known as a greater ecosystem. Examples: the southern, west slope of the Sierra Nevada is one bioregion; Death Valley is another.

bitter tonic. A substance that stimulates digestive functions, first in the mouth and later in the stomach and liver. Bitter tonics are traditionally used to aid digestion.

bracts. Modified or reduced leaflets usually associated with the flower of a plant; often located beneath the petals.

cambium. The cell layer between the wood and bark of trees and shrubs. The cambium is where outward trunk or limb growth takes place, and where a continuity of biochemical and cellular activity is maintained and carried to all parts of a tree.

candida. A yeastlike genus of fungi that inhabit the vagina or digestive tract, or both, which under certain conditions may cause candidiasis, an acute or chronic yeast infection.

carcinogen, carcinogenic. A substance or agent that promotes the formation or growth of cancer.

cardiac tonic. Capable of strengthening the heart muscle or stimulating heartbeat, or both, in a manner beneficial to body functions.

carminative. A substance that aids in the expulsion of gas from the digestive tract, having a "carminative action" in the body.

catarrh. The excessive secretion of thick phlegm or mucus from inflamed mucous membranes. See *anticatarrhal.*

cathartic. A synonym for *laxative.* This term generally is used to describe a strong laxative.

catkin. An elongated, often fuzzy, conelike flower usually lacking any distinguishable petals or sepals. Catkins are characteristic of the Poplar and Willow plant families. (See willow photo.)

chlorophyll. Any of an assortment of green pigments found in plants. Chlorophyll enables plants to photosynthesize.

cholagogue. Refers to substances that stimulate bile production in the liver.

circumboreal. The portion of the earth that includes the northern one-third of the planet.

coagulant. A substance capable of promoting blood clotting, converting blood from a liquid to a semisolid state.

compound. In reference to leaves, a compound leaf is composed of multiple, smaller leaf segments, which often are pinnately arranged pairs of leaflets. (See angelica, elderberry, and yarrow photos.)

concentric. Having a common center; circular. (See angelica and lomatium photos for concentric flower clusters.)

conifers. Trees and shrubs that bear their flowers and fruits in the form of scaly, conelike structures; includes all members of the fir, pine, and cypress families.

constituent. A single element or a compound ingredient that is part of a whole. A medicinal constituent in a plant is an element or compound that makes the plant medicinally useful.

corm. A swollen, nutlike structure on the root systems of various plants.

corolla. The collective petals, or rays, of a flower.

counterirritant. An irritant that distracts attention away from another irritant. Usually applied externally. A deep-heating, Mentholatum muscle ointment is a counterirritant to the discomfort of aching muscles.

deciduous. Plants that lose their leaves once a year, at the end of the growing season.

decoction. An herbal preparation made by simmering plant material in water until maximum extraction of active constituents is achieved. This process is usually used for roots, barks, and seeds that are not water soluble enough for use in simple infusions (teas).

demulcent. A substance that provides a protective coating and is soothing to irritated tissues in the body.

dermatitis. Inflammation of the skin.

detritus. Loose fragments on the forest floor that result from the disintegration of rocks and forest debris.

diaphoretic. Capable of stimulating perspiration.

digestive tonic. A substance that aids digestion.

diuresis. The process by which the body eliminates waste and excesses through kidney function and subsequent urination.

diuretic. A substance that stimulates diuresis.

duff. Thick mats of detritus on the forest floor.

earth-conscious. Acting with respect to the natural needs, sustainability, and design of one's environment.

earth regenerator. A plant or other organism that helps to repair damaged habitat or soil structure.

ecoherbalist. A person who studies, uses, and cares for wild medicinal plants at a level of understanding and respect that gives precedence to the natural and holistic needs, cycles, and sustainability of the ecosystem.

ecosystem. The interdependent interaction between an ecological community and its environment.

elliptical. A longer-than-wide, oval-like shape with opposite ends that are equal in diameter. Not egg shaped, but like a flattened circle.

emetic. A substance that induces vomiting.

emmenagogue. A substance that promotes menstruation.

emollient. A substance that soothes, protects, and softens the skin. The external counterpart to a demulcent.

evergreen. Any of a wide variety of plants (not just conifers) that retain most or all of their foliage through the winter months.

expectorant. A substance that helps to expel mucus from the upper respiratory tract.

filament. In a flower, the anther-bearing stalk of a stamen.

flavonoid. A chemical compound found in various forms in several plants. Flavonoid constituents are responsible for a wide range of medicinal actions and are generally responsible for the pigmentation of various red, yellow, or purple fruits. Also known as a bioflavonoid.

fruit. The seed-bearing structure of a plant.

glabrous. Lacking hairs. A characteristic of certain plant stems and leaves.

hemispherical. A shape that represents an equally divided half of a sphere.

hemostatic. Refers to the stoppage of bleeding. Most herbal hemostatic substances work by astringent actions.

hepatic. A general term referring to medicinal action on the liver.

homeopathy. A modality of medicine based on the theory that diseases can be cured by administering very minute doses of drugs that in a healthy person would produce symptoms similar to those of the disease.

homeostasis. A maintained state of health where all checks and balances between interdependent elements of mind, body, and spirit are functioning harmoniously.

hypertension. High blood pressure.

hyperthyroidism. Overactive functioning of the thyroid gland.

hypotension. Low blood pressure.

hypotensive. Capable of reducing blood pressure.

hypothyroidism. Depressed thyroid function.

immune modulator. Often used synonymously with the term *immunostimulant,* this term is earning its own place in the herbalist's vocabulary as we begin to understand more about how certain immune-supporting herbs work in the body. An immune modulator is an herb that acts to help the immune system by optimizing its ability to adjust to stressful circumstances.

immunostimulant. An herb that strengthens the body's resistance to infection by stimulating and increasing immune system responses. In herbal medicine, this term specifically refers to the medicinal support of infection-fighting antibodies in the bloodstream and overall tonification of the lymph system.

inflorescence. A flower or cluster of flowers.

infusion. A preparation made by pouring boiling water over herbs and allowing it to steep; a tea.

lanceolate. Lance-shaped; widest at the base, with sides tapering to a point. (See photos of pipsissewa, which has narrowly lanceolate leaves, and arnica, which has broadly lanceolate leaves.)

laxative. A substance that increases the frequency of bowel evacuation, generally through softening of fecal matter.

linear. Long and narrow. Referring to leaf characteristics, a linear leaf is too narrow to be considered "narrowly lanceolate," but instead resembles a blade of grass.

lobed. Referring to leaf characteristics, a lobed leaf has margins (outer edges) that are deeply indented in two or more places, but not as deeply as a palmate leaf. Maple leaves are deeply lobed. (See alumroot and cow parsnip photos.)

lymph system. The system of the body responsible for the cleansing of tissues at the surface to cellular level and the production of various antibodies and white blood cells.

lymphatic. Refers to the lymph system. A lymphatic tonic strengthens the function of the lymph system.

medicinal action. Any of a variety of terms used to describe the effect an herb or other substance has on or in the body that may be considered therapeutic.

microbe. A microscopic organism, including various bacteria, viruses, fungi, and protozoa (such as *Giardia*).

microcosm. A tiny, often microscopic segment of an ecosystem represented by a specialized community of organisms; for example, the interrelated community of organisms in a patch of moss on a sheer rock cliff.

microecosystem. A small segment of an ecosystem. The relationship between the microorganisms on a dead tree and their environment is an example of a microecosystem.

mucilaginous. Containing mucilage, a sticky-oily substance often used in herbal medicine to soothe and protect irritated tissues. Mucilaginous herbs are generally used as emollients or demulcents.

nephritic. Of or relating to the kidneys.

nervine. Refers to substances that are tonic to the nervous system.

noxious weed. A term used in reference to plant species we do not like and want to eradicate.

ointment. Somewhere between a liquid extract and a salve. An ointment is a thickened herbal extract that is too loose to hold its form without a container.

open-pollinated. Plant species that have not been genetically altered (nonhybridized) and are capable of reproducing in true form from generation to generation.

opposite. Referring to leaf characteristics, opposite leaves are arranged directly across from each other at regular intervals along the stem of the plant. (See self-heal photo.)

ovate. Oval-shaped. (See hawthorn photo.)

palmate. A shape that resembles the human hand with fingers extended. A palmate leaf has margins deeply indented nearly to its base.

pathogenic microbes. Microscopic organisms that act negatively on or in the body; harmful, infectious bacteria, fungi, and viruses are pathogenic microbes.

peduncle. The stemlike structure that holds the fruit or flowers of many plants. Also commonly known as a pedicel. (See catnip photo.)

perennial. A plant that returns from its rootstock year after year. Perennials reproduce by seed and root reproduction.

petals. The bractlike inner segments of a flower; usually the most colorful part.

petiole. The part of a plant generally viewed as a leaf stem; it joins the leaf to the stem or root crown of the plant. (See photos of arrowleaf balsamroot, which has proportionately long petioles, and self-heal, which has short or nearly absent petioles.)

pH. A numerical measurement of acidity or alkalinity. Relating to soil, a pH level of 7.0 is regarded as neutral.

photosynthesis. The process by which plants convert solar energy to carbohydrates.

phytopharmaceutical. A scientifically engineered plant medicine; a "plant drug."

pinnate. A compound leaf pattern where leaflets are arranged in opposing pairs along two sides of an axis. (See angelica, elderberry, and valerian photos.)

pollinator attractor. A plant that is particularly attractive to insects and other organisms vital to pollination between plants.

poultice. An herbal preparation made by mashing plant materials with a liquid (usually water) to form a wet paste.

purgative. An extremely, perhaps violently, strong laxative with uncontrollable effect. Purgatives generally cause abdominal cramping and near-incontinent conditions. They are usually reserved by herbalists for use only in dire circumstances.

raceme. A flower cluster in which individual flowers have stalks and are attached to an elongated axis or stem. (See false Solomon's seal photo.)

ray. The extended, bladelike petal of a ray flower. (See photos of arnica and arrowleaf balsamroot, both with yellow rays.)

refrigerant. Capable of cooling the body temperature, usually by means of perspiration.

rhizome. An underground stem that extends itself horizontally. Crabgrass is an excellent example of a rhizomatous plant.

riparian habitat. An ecosystem in proximity to a consistent source of water (such as floodplains, stream banks, lakeshores, and marshes).

rosette. Referring to leaves, those which emerge in an overlay pattern resembling the shape of a rose flower. (See mullein photo.)

rubifacient. A substance that reddens and heats the skin when applied topically. A mustard pack is a classic example of a rubifacient.

salve. A preparation consisting of an herbal extract (usually an infusion or decoction) that has been thickened to the consistency of butter.

saponin. A glycoside (soaplike) plant compound. Present in many species of plants, saponins are characterized by their soaplike nature. Although many types of saponin have medicinally useful properties, many may also be irritating to the digestive tract and may cause toxic reactions if ingested.

sedative. An action that calms the nerves to help a person relax.

sepal. A modified leaf (usually green) that encloses a flower bud. Some plants have sepals that are more conspicuous than their flowers; for example, coptis (with green sepals) and bunchberry (with white sepals).

simple. Referring to leaf characteristics, a simple leaf has margins (outer edges) that are void of any serrations, divisions, or lobes. Take it literally: a simple leaf is a basic leaf.

sitz bath. A method of bathing where only the pelvic area of the body is immersed. Astringent herbs are sometimes used in sitz bath therapies for treatment of hemorrhoids, postpartum swelling, and other types of inflammation below the waist.

stamen. The male, pollen-bearing part of a flower, which collectively consists of an anther and a filament.

stigma. The female, pollen-receptive part of a flower, which is usually at the uppermost part of the flower center.

stimulant. A general term used to describe the increase of functional activity.

stratification. A germination process by which a seed must be subjected to a prolonged period of cold (often freezing) temperatures and moisture to break its dormancy.

Streptococcus. A genus of gram-positive, spherical bacteria in microscopic chains. Most forms of *Streptococcus* are normally present and harmless in the body, while others, such as *S. pneumoniae* (bacterial pneumonia), are potentially deadly.

styptic. An agent that causes bleeding to stop by making tissues contract rapidly; essentially the same as an astringent.

subalpine. A general term for the ecological zone that lies below timberline in montane areas. Exactly where the subalpine zone begins and ends is often unclear and arguable among botanists, who generally differentiate subalpine from alpine zones by marker species, elevation, and climatic variances. The subalpine zone ends where dense stands of conifers become sparse or nonexistent at higher elevations.

symbiosis. A relationship in which two dissimilar organisms live together for mutual benefit.

terminate. Referring to flowers that are the absolute end-tips of plant stems. Also referred to as *terminal.* (See angelica, arrowleaf balsamroot, bee balm, and self-heal photos.)

tincture. An herbal preparation made by soaking plant material in a liquid solvent (called a "menstruum") to extract active medicinal constituents. Commonly referred to as herbal extracts, tinctures may be made from menstruums of alcohol, glycerin, or vinegar. The type of menstruum used for a particular herb depends on the chemical or physical structure of the herb, or both. Alcohol-based menstruums usually yield the strongest herb tinctures and have the longest shelf life.

tinnitus. Ringing or buzzing in the ear that cannot be attributed to external stimulus. The causes of chronic tinnitus are subject to debate; many people believe microbial infection, circulatory problems, or physical defects of the inner ear may lead the list of contributing factors.

tonic. A general term for a nourishing substance that invigorates and increases the tone and strength of tissues and improves the function of one or more body systems.

umbel. Referring to flowers, *umbel* means umbrella-shaped. True umbels consist of tiny florets, each extending an equal distance from a common point to form dense clusters (see angelica and cow parsnip photos). I have used the terms *umbel* and *umbel-like* loosely, to include flowers that appear umbrella-shaped regardless of the "true" criteria.

uterine stimulant. Capable of stimulating contraction of the uterus.

uterotonic. Tonic to the uterus.

vasoconstrictor. Capable of tightening the walls of blood vessels. Opposite of *vasodilator*.

vasodilator. Capable of dilating or widening blood vessels. Opposite of *vasoconstrictor*.

vermifuge. Expels worms.

vulnerary. Refers to various actions that promote the healing of wounds.

FLOWER ANATOMY

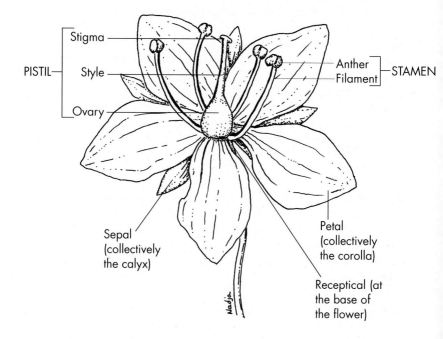

BIBLIOGRAPHY AND RECOMMENDED READING

❦

Belzer, Thomas J. *Roadside Plants of Southern California.* Missoula, Mont.: Mountain Press, 1984.

Bergner, Paul. *The Healing Power of Echinacea & Goldenseal.* Rocklin, Calif.: Prima Publishing, 1997.

Buchman, Dian Dincin. *Herbal Medicine.* New York: Gramercy Publishing, 1980.

Castleman, Michael. *The Healing Herbs.* Emmaus, Pa.: Rodale Press, 1991.

Dawson, Ronald L. *Nature Bound: Pocket Field Guide.* Boise, Idaho: OMNIgraphics, 1985.

Densmore, Frances. *How Indians Use Wild Plants for Food, Medicine, and Crafts.* New York: Dover, 1974.

Dorn, Robert D. *Vascular Plants of Montana.* Cheyenne, Wyo.: Mountain West Publishing, 1984.

Duke, James A., and Steven Foster. *Eastern/Central Medicinal Plants.* Peterson Field Guide Series. Boston: Houghton Mifflin, 1990.

Foster, Steven. *Herbal Renaissance.* Layton, Utah: Gibbs Smith, 1993.

Given, David R. *Principles and Practice of Plant Conservation.* Portland, Oreg.: Timber Press, 1994. (A specific source of information on ethical wildcrafting.)

Gladstar, Rosemary. *Herbal Healing for Women.* New York: Simon & Schuster, 1993.

Green, James. *The Male Herbal.* Freedom, Calif.: The Crossing Press, 1991.

Grieve, Mrs. M. *A Modern Herbal.* Vols. 1 and 2. New York: Dover, 1971.

Harrington, H. D. *Edible Native Plants of the Rocky Mountains.* Albuquerque: University of New Mexico Press, 1967.

Hart, Jeff. *Montana: Native Plants and Early Peoples.* Helena: Montana Historical Society, 1976.

Hitchcock, Leo C., and Arthur Cronquist. *Flora of the Pacific Northwest.* Seattle: University of Washington Press, 1973.

Hoffmann, David. *An Elder's Herbal.* Rochester, Vt.: Healing Arts Press, 1993.

———. *The New Holistic Herbal.* Rockport, Mass.: Element Books, 1990.

Kindscher, Kelly. *Medicinal Wild Plants of the Prairie.* Lawrence: University Press of Kansas, 1992.

Little, Elbert L. *The Audubon Society Guide to North American Trees, Western Region.* New York: Knopf, 1988.

Lust, John. *The Herb Book.* 20th ed. New York: Bantam, 1987.

Mabey, Richard. *The New Age Herbalist.* New York: Collier Books, 1988.

Mills, Simon Y. *Out of the Earth.* New York: Viking, 1991.

Moore, Michael. *Medicinal Plants of the Desert and Canyon West.* Santa Fe: Museum of New Mexico Press, 1989.

———. *Medicinal Plants of the Mountain West.* 6th ed. Santa Fe: Museum of New Mexico Press, 1988.

———. *Medicinal Plants of the Pacific West.* Santa Fe, N.Mex.: Red Crane Books, 1993.

Mowrey, Daniel B. *Herbal Tonic Therapies.* New Canaan, Conn.: Keats Publishing, 1993.

Newall, Carol A., Linda A. Anderson, and J. David Phillipson. *Herbal Medicines: A Guide for Health-Care Professionals.* London, England: Pharmaceutical Press, 1996.

Rocky Mountain Herbalist Coalition. *Direct Marketing Registry: Ethical Wildcrafters and Organic Growers of Medicinal Herbs.* Boulder, Colo.: Rocky Mountain Herbalist Coalition. (An annually published specific source of information on ethical wildcrafting.)

Schofield, Janice J. *Discovering Wild Plants: Alaska, Canada, the Northwest.* Anchorage: Alaska Northwest Books, 1989.

Schultes, Richard Evans. *Medicines from the Earth.* Rev. ed. New York: McGraw-Hill, 1983.

Spellenberg, Richard. *The Audubon Society Field Guide to North American Wildflowers.* New York: Knopf, 1979.

Strickler, Dee. *Forest Wildflowers.* Columbia Falls, Mont.: Flower Press, 1988.

Taylor, Ronald J. *Northwest Weeds.* Missoula, Mont.: Mountain Press, 1990.

———. *Sagebrush Country.* Missoula, Mont.: Mountain Press, 1992.

Taylor, Ronald J., and George W. Douglas. *Mountain Plants of the Pacific Northwest.* Missoula, Mont.: Mountain Press, 1995.

Thie, Krista. *A Plant Lover's Guide to Wildcrafting.* White Salmon, Wash.: Longevity Herb Press, 1991. (A specific source of information on ethical wildcrafting.)

Tilford, Gregory L. *The EcoHerbalist's Fieldbook: Wildcrafting in the Mountain West.* Conner, Mont.: Mountain Weed Publishing, 1993.

———. *Edible and Medicinal Plants of the West.* Missoula, Mont.: Mountain Press, 1997.

Weiss, Rudolf Fritz. *Herbal Medicine.* Beaconsfield, England: Beaconsfield Publishers, 1988.

Werbach, Melvyn R., M.D., and Michael T. Murray, N.D. *Botanical Influences on Illness: A Sourcebook of Clinical Research.* Tarzana, Calif.: Third Line Press, 1994.

Whitson, Tom D. *Weeds of the West.* Rev. ed. Newark, Calif.: Western Society of Weed Science, 1992.

Willard, Terry. *Edible and Medicinal Plants of the Rocky Mountains and Neighbouring Territories.* Calgary, Alberta, Can.: Wild Rose, 1992.

INDEX

∾

Juniperis occidentalis, 135
Juniperis osteosperma, 135–36
Juniperis species, 135–38

kidney: disease, 36, 135; inflammation, 158; irritation, 36, 135, 196; stones, 109
kinnickinnick, 196. *See also* uva-ursi
klammath weed, 184
knapweed, 190

Labiatae, 52, 64, 71, 126, 181, 188
lacerations, 87, 109, 132
Lactuca species, 90
lappa, 67
lavender, 186
laxatives, 74, 90, 106, 116, 160, 171, 213
lead, 134
lecithin, xi, 90
Leguminosae, 174
lemon balm, 186
Leonuris cardiaca, 62, 64
lettuces, wild, 90
licorice root, 68, 106
ligament injuries, 45, 128. *See also* connective tissue
Ligusticum canbyi, 5
Liliaceae, 102
lily family, 102
lily of the valley, false, 11
lily of the valley, wild, 102
lion's tooth, 90
lip salve, 166
liver: detoxification of, 174; dysfunction, 68; stimulant, xi, 90
logging, commercial, 160
lomatium, 126, 139–42
lomatium, fern-leafed, 139
Lomatium dissectum, 139–42
lotion, 74
lupine, 87
Lycopus americanus, 64
lymphatic tonic, 78
lymph system, 45, 77, 210

madder family, 77. *See also* cleavers
Maianthenum dilatatum, 11
male tonic, 172

manganese, 146
manzanita, 196
marrubio, 126
Marrubium vulgare, 64, 126–27
marshmallow, 50, 54, 87, 109, 126, 129, 160, 144, 177, 196, 214
masterwort, American, 83
Matricaria matricarioides, 154–56. *See also* pineapple weed
mayblossom, 123
meadowsweet, 46, 208
medicine, big, 139
Mellisa officinalis, 186
menopause, 42, 56
menses, painful delayed, 56. *See also* emmenogogues; uterine disorders
menstrual cramping, 42, 60
menstrual flow regulator, 146
menstrual stimulant. *See* emmenogogues
menstruum, 24
migraine headaches, 80
milfoil, 210
mint family, 181, 188
mint, 52, 64, 71, 85, 126
miscarriage, 60
mites, 144
Monarda fistulosa, 52–55
Monarda species, 52–55
mood elevators, 104
Moore, Michael, 64, 96, 190
moth repellant, 211
motherwort, 62, 64
mountain ash, 98
mouth: infections, 96; inflammations, 112, 166; irritations, 74, 206; ulcerations, 104, 182
mouthwash, 36, 52, 177
mucilaginous herbs, 36, 102, 160
mucous membranes, 63, 80
mule's ears, 49
mullein, 50, 54, 104, 109, 126, 132, 141, 143–45, 211
muscle pain, 137
muscle tissues, smooth, 124
mustard family, 192

Nepeta cataria, 71–73
nephritics, 157, 168

Gregory L. Tilford —Mary Tilford photo

ABOUT THE AUTHORS

An internationally renowned herbalist and naturalist, **Gregory L. Tilford** travels North America teaching about herbs and herbalism and lecturing at North America's top institutes of herbal study. Tilford is a member of the American Herbalist Guild and the Natural Pet Products Association and author of *Edible and Medicinal Plants of the West* (Mountain Press, 1997). Mr. Tilford and his wife, Mary, own and operate Animals' Apawthecary, makers of herbal health care products for dogs and cats, in Conner, Montana.

Rosemary Gladstar, author of *Herbal Healing for Women,* is one of the best known and most respected teachers of herbalists in the world. She cofounded the California School of Herbal Studies, founded United Plant Savers, and advises many of the world's most respected herb research organizations and product manufacturers. Ms. Gladstar currently runs Sage Mountain Herbal Retreat Center and Botanical Sanctuary in East Barre, Vermont.

ABOUT THE ILLUSTRATOR

Nadja Cech Lindley illustrated her first catalog at the age of twelve, and her artwork has graced the pages of *Herbalgram, The Herbalist, The Pharmer's Almanac,* and Horizon Herbs publications. Now a graduate student in chemistry at the University of New Mexico, Ms. Lindley continues to create botanical art, drawing inspiration from the natural beauty of medicinal plants.

We encourage you to patronize your local bookstores. Most stores will order any title that they do not stock. You may also order directly from Mountain Press by mail, using the order form provided below or by calling our toll-free number and using your Visa or MasterCard. We will gladly send you a complete catalog upon request.

Some other Natural History titles of interest:

____Alpine Wildflowers of the Rocky Mountains	$14.00
____Beachcombing the Atlantic Coast	$15.00
____Birds of the Central Rockies	$14.00
____Birds of the Northern Rockies	$12.00
____Birds of the Pacific Northwest Mountains	$14.00
____Coastal Wildflowers of the Pacific Northwest	$14.00
____Desert Wildflowers of North America	$24.00
____Edible and Medicinal Plants of the West	$21.00
____From Earth to Herbalist:An Earth-Conscious Guide to Medicinal Plants	$21.00
____Graced by Pines The Ponderosa Pine in the American West	$10.00
____A Guide to Rock Art Sites Southern California and Southern Nevada	$20.00
____Hollows, Peepers, and Highlanders An Appalachian Mountain Ecology	$14.00
____An Introduction to Northern California Birds	$14.00
____An Introduction to Southern California Birds	$14.00
____The Lochsa Story Land Ethics in the Bitterroot Mountains	$20.00
____Mammals of the Central Rockies	$14.00
____Mammals of the Northern Rockies	$12.00
____Mountain Plants of the Pacific Northwest	$25.00
____New England's Mountain Flowers	$17.00
____Northwest Weeds The Ugly and Beautiful Villains of Fields, Gardens, and Roadsides	$14.00
____OWLS Whoo are they?	$12.00
____Plants of Waterton-Glacier National Parks and the Northern Rockies	$12.00
____Roadside Plants of Southern California	$15.00
____Sagebrush Country A Wildflower Sanctuary	$14.00
____Watchable Birds of the Rocky Mountains	$14.00
____Watchable Birds of the Southwest	$14.00

Please include $3.00 per order to cover shipping and handling.

Send the books marked above. I enclose $_____

Name_____

Address_____

City_____State_____Zip_____

☐ Payment enclosed (check or money order in U.S. funds)

Bill my: ☐ VISA ☐ MasterCard Expiration Date:_____

Card No._____

Signature _____

Mountain Press Publishing Company
P.O. Box 2399 • Missoula, Montana 59806
Order Toll Free 1-800-234-5308 with your Visa or MasterCard
E-MAIL: mtnpress@montana.com • WEBSITE: www.mtnpress.com